Spend Analysis and Specification Development

Using Failure Interpretation

Spend Analysis and Specification Development Using Failure Interpretation

Michael D. Holloway

CRC Press is an imprint of the
Taylor & Francis Group, an **informa** business

CRC Press
Taylor & Francis Group
6000 Broken Sound Parkway NW, Suite 300
Boca Raton, FL 33487-2742

Printed in the United States of America on acid-free paper
Version Date: 20110726

International Standard Book Number: 978-1-4398-5107-4 (Hardback)

Library of Congress Cataloging-in-Publication Data

Holloway, Michael D.
Spend analysis and specification development using failure interpretation / Michael D. Holloway.
p. cm.
Includes bibliographical references and index.
ISBN 978-1-4398-5107-4 (hbk. : alk. paper)
1. Production management. 2. Costs, Industrial--Evaluation. 3. Cost control. 4. Production control. 5. Industrial procurement. I. Title.

TS155.H6454 2011
658.7'2--dc22 2011014663

Visit the Taylor & Francis Web site at
http://www.taylorandfrancis.com

and the CRC Press Web site at
http://www.crcpress.com

For my mother and father—the two finest people there will ever be.

Contents

Foreword

A while ago, I had the great pleasure of presenting a paper to the American Iron and Steel Engineers Annual Conference. My paper was about increasing the reliability of continuous casting operations by way of improved lubrication technology. The audience was a mature lot in both experience and age. I shared insight on how operations could actually save money through a better selection process of the lubricants used on the bearings for the operation. I shared the technology and the selection process as well as case studies and documented savings. I actually felt as though I knew something about how the steel industry worked and that a few of my ideas might just make a real difference. It was not long after my talk that I was brought back down to reality and "schooled," so to speak. An elderly gentleman approached me, stuck out his hand, and said "Son, you did a fine job. I really enjoyed your talk today." We shook hands, and I thanked him and proceeded to share with him how I loved visiting various plants and learning about how things operate. "You seem like a fine young man, smart. Now don't take offense, but I don't think you know jack about how important grease is in a bearing." I did not take offense. Actually, I was intrigued. Often after I give a presentation or a seminar, one or two participants will try to challenge me on my experience or knowledge. I enjoy these discussions; it is where I learn the most. "Okay sir, I'm game. I've been doing this long enough to spot a man with a story, and I will bet this story is going to change my perspective on something, isn't it?" I said. I had no idea that what this man was going to share with me was going to change how I looked at the world forever.

He began:

> Many years ago I worked right down on a continuous casting line for a pretty darn big mill. We supplied pretty much all the structural steel for bridges and buildings in North America as well as around the world. Now as you know, on the continuous casting line the roller bearings are helping form the steel and let me tell you, they get punished. The only thing helping them is the grease that is being pumped through. Well, corporate decided to hire some MBA-type consultants to come in and help us save some money. I had a feeling that one of the suits in the upper

> office went to business school with a few of them and had to pay back a favor or something. Anyway, these guys come in and took a look around for awhile. I guess they were trying to figure out how they were going to save us some money. They took a look at our "consumables"—you know, the stuff a maintenance and repair department might go through on a regular basis. Well they settled on grease. They saw how much grease we were going through and how much we were paying and decided to call up some bigwig at (name omitted) to supply a cheaper grease. Stuff came straight from the company and not even through a regular distributor. I figured they made another deal with someone else they knew from business school, the old boy network you know. So we start using this grease. Now mind you, the stuff we were using before was more expensive but the company that we got it from (name omitted) spent some time with us down on the floor figuring out what we really needed. Well we start using this new grease and it wasn't maybe 2 weeks before our first bearing failure. We hadn't had a bearing failure in years! Management said it was just one of those things. We told them the new grease was terrible but they wouldn't listen. They were saving big money on their consumables, heck this new grease was about four times less expensive than what we were using. Now when you get a bearing failure it pretty much shuts the line down for several hours, takes a team of at least six men to fix it. You know how much one of those bearings costs? Thousands. You know how much downtime costs at a mill? Many thousands. A few weeks later we had another bearing fail. Seized right up. Then a few more. This idea they had to save money was costing them millions. Well the geniuses up on Mahogany row couldn't see that they made a huge mistake or maybe they could but just wanted to save face or something so instead of going back to what we were using they decided to cut back on overtime for the maintenance guys and some of production as well. Let me tell you, that didn't sit well. With money being lost in line slowdowns, something had to be done with a fast impact.

I thought I knew where he was going with this story. I saw it all too often. The most expensive direct cost in an organization always seems to be people. Failure was taking on a whole new meaning. Reduce people or the hours they worked, and you can reduce the money bleeding out of the organization. But at what cost?

"Sounds like a typical management-type blunder," I said.

He looked at me with steely blue eyes. The eyes of an old man tired of the mistakes he had witnessed and tired of the waste of time, money, and effort. His thousand-yard stare was one of a man trying to look beyond all the foolishness and toward something that made sense—something that meant something. "Son, you have no idea. Now let me ask you something, what would you do if your boss cut your pay by, oh let's say 30% or so?"

I thought for a second or two and said, "I would have a big problem with that, and I would most likely look for another job."

He responded, "And I would bet you would have no problem finding one either. You might have to move but a young guy like you wouldn't mind that much. But let me tell you, it wasn't that easy. You see this mill was all we had. This was where our fathers, grandfathers, and even great-grandfathers had worked. Most of us didn't even graduate from high school. We didn't have to. Let me tell you, when you take away a man's income you take away part of who he is and if he can't do anything about it he starts down a road of self-destruction. Some guys drink, others gamble. Other guys just get back problems, ulcers, heart problems, or just get plain depressed and decide to give up. That's what began to happen."

At this point, a few other men joined in the discussion, telling of guys who would miss work for days trying to sleep off a 3-day drunk or a few who just up and disappeared and still others who got into all sorts of trouble with the law because they would fight with their wives or friends or themselves.

At this point, I was sitting down listening intently to what he had to say. I felt as if I was listening to my grandfather tell me about the trench warfare he experienced during World War I or the stories my own father would tell me about the Battle of the Bulge, which he fought in. He continued, "It wasn't long before upper management didn't like all the missed time. They really didn't like all the production quotas that were being missed either and I'm sure the stockholders didn't like the dip in the dividend checks. Of course the union had it under control. Or so they thought. You would have thought that with the cutbacks in overtime the union would have gone crazy, but I guess concessions were made and I'm sure palms were greased. It was the beginning of a cancer that would spread."

As the man spoke, it occurred to me that his story took on a much more personal bend than what I was used to. I had done countless cost studies as well as reliability proposals, but it never really occurred to me that there were always people behind those numbers, living souls with families behind each wrench, wheel, or control panel.

He went on, "Upper management decided that if you were gonna miss days you better have a good excuse so you needed a note from your doctor. Well, I don't know what they were thinking by having the men get notes from there docs—heck was that supposed to scare them? These guys poured iron for a living, they weren't afraid of anything let alone some squirrelly doctor. These guys started going to the doctors all the time for anything and everything. It took about two and a half months before the corporate guys were going crazy over all the additional costs they had to pay out in the ways of increased health insurance. It didn't take a New York second for them to go and cut back on the medical benefits. Now this didn't bode well at all with the union. There was an emergency union meeting called—typical smoke-filled, dirty hall with all sorts of yelling going on. It was quickly decided what we had to do."

I knew what they did. I remembered it very well. "You worked for Bethlehem Steel didn't you?" I said.

"Yup, many of us did," he said with a smile.

"Your union called a strike, pretty much destroyed Bethlehem. The labor negotiations are the stuff of graduate business school legends!" I said.

"Let me tell you something son, what I'm telling you can't be found in any of those books you had in school. Heck, most of that crap is written by the idiots that caused the mess to begin with!"

I didn't take exception to what he was saying. He was right. There was tons of theoretical stuff in textbooks that frankly just wasn't applicable to the real world. There is a place for all that, just not on the "floor."

"This strike couldn't have happened at a worse time. Just as we were squabbling, China starts building steel mills and supplying all the customers we couldn't due to all the line shutdowns, union issues, and increased costs."

The growth of the Chinese steel industry was explosive. It can also be concluded that when you build a primary business like steel, then other allied industries start to crop up to not only support that industry but also to take advantage of the local supply of raw materials. In China, some cold-roll steel as well as durable goods were being made at a lower product price and a higher yield than the American counterparts. The only thing China needed was a way to get all those goods cheaply distributed to the United States. May I introduce you to Wal-Mart?

Was that old man waxing poetic, spinning tales about days gone by, trying to school a young punk? Maybe he was. Was the world economy turned upside down because of a poor decision on the grease used to lubricate a bearing? It may have been, but it should not be a surprise to any of us that a failure of such a small component like a bearing is a type of cancer that runs through a company. And sometimes cancer can be caused by the littlest of things. *Cancer* is a ubiquitous term for something that comes about through:

- Genetic predisposition
- Environmental exposure
- Errors in DNA repair
- Accumulated genetic errors in somatic cells
- Ineffective immunosurveillance

Failure is also a broad term defined as the state or condition of not meeting a desirable or intended objective event. Some will say that the opposite of failure is success, yet each refers to an event that takes place. Positive or negative, it is an event. Failure is never something someone sets out to do. The objective is to "do" and be successful, but what makes up this "do?" This thing we know as an "event," why does it occur? Before we can even begin to understand success and failure, we must understand what "being" is, where it comes from, and eventually what the ramifications will be. For those of you interested in reading more on this subject, look toward *ontology* (the philosophical study of the nature of being, existence, or reality) as well as *metaphysics* (a branch of philosophy concerned with explaining the fundamental nature of being and the world). There are those who enjoy reading about such topics, while others need it all broken down into really easy parts. Richard Feynman wrote a book, *Six Easy Pieces*, that explains the essentials of physics. After reading his book, the concepts of energy, gravity, and quantum behavior are within easy reach of even a nonscientific person. As I began to try to understand why things occurred in various industries, I had to develop my own "easy pieces" in order to understand why events happen, specifically failure.

It is not always one event plus another event that will result in a third event, also known as the additive principle (or $1 + 1 = 2$, for those of you inclined to understand the world in terms of equations). Sometimes it can be

- *Synergistic*: $1 + 1 \neq 2, >2$, where two or more events combined have a greater or significantly different effect than the sum of effects of the things applied separately. Think of the example of teamwork, which can produce an overall better result than if each person worked on the same goal individually.
- *Potentialistic*: $0 + 1 \neq 1, >1$, where an event becomes greater than itself without an additive or synergistic event occurring. The event becomes a catalyst for a larger event change.
- *Antagonistic*: $1 + 1 \neq 2 = 0 < 2$, where two or more events have an overall effect that is less than the sum of their individual effects.

Understanding the "how" of an event is valuable; understanding the "why" of an event is priceless. If you can determine why something went wrong and figure out what is required so that the next thing you buy does not have the same event, you just might save yourself and your company a lot more than hitting your quarterly operating budget. What you buy influences your ability to make a profit or "go the way of all flesh."

When we consider that the biggest machines that do the most work are made up of smaller machines that are made up of components, it becomes obvious that when a large machine breaks, it is normally due to a small component or components acting synergistically, potentialistically, or antagonistically. How many times have we taken our vehicles to the shop only to find out that we have to pay big money for a small part? Or worse, how often do we hear about major recalls of vehicles due to a small but critical part malfunction? Too often.

This book presents a method by which you can lower your operational costs and increase productivity by establishing a performance-based procurement specification on the components that are a cancer to your facility, plant, mill, or mine. In order to develop the specification, it is vital that you understand what you are spending the most money on, also known as a *spend analysis*. This is covered in the book. It will also require a basic understanding of influential components and materials' physical properties and the ability to engage in some fundamental failure analysis, which are covered in this book. Finally, to get the most out of these methods, it is necessary to work closely with the men and women who spend the majority of their lives with this equipment. Words to the wise, do not try to take on too many items at first. It is better to work thoroughly and completely than to rush through and misapply these principles. You may even be able to use some of these methods in your personal purchasing practices. But keep in mind, do not apply these principles to buying your wife or girlfriend a birthday present. For those purchases, the more money you spend, the greater the reward.

Acknowledgments

There are many reasons why people write books, including money, fame, and respect. Those reasons are all plausible, but the reason I wrote this particular piece of work is because I could not find anything else out there that covered this topic in this way. The more people who I talked to about the concepts of spending analysis and specification development using failure as a means of understanding it, the more it became obvious that it was a great topic, both timely and timeless, and it was something many people from many professions and educational paths could relate to and use.

When I first told my family that I was going to write a book, they thought it was going to be a storybook complete with pictures. Well, there are plenty of stories and even a few diagrams. I have to thank them for their constant encouragement and desire to see this project through. Thank you to Lisa, Max, and Emma, and a special thanks to my father, Leon Holloway, who never missed an opportunity to ask about the progress and find words of encouragement. He is the most amazing man I will ever know. I also would like to thank my brother Mark and his wife Lori for helping with the flow and general edits, and my brother Chris for his insight and suggestions. I would like to acknowledge two men who have provided leadership, compassion, intellectual freedom, and focus, Jim Rodgers and Bill Ramey. Although they could not be farther apart on the personality spectra, they are the finest industry leaders with whom I have ever had the pleasure of working with. I would like to thank the Levy family for building one of the finest companies in the world and allowing me to work for them while completing this manuscript. I would also like to thank several other people who have provided insight and candor while I worked on this piece: Captain Glenn Kallgren (U.S. Army, Retired), Captain Paul Morgan (U.S. Navy, Retired), Aimee Mullins, Fran Christopher, Kevin Tate, David Myers, and Suzanne Remore, as well as Linda Leggio. I would be remiss if I did not acknowledge Janet Zito for her continuous influence—35 years later and still not forgotten. Thank you for not stopping. I would also like to thank my newest friend in this wonderful world of publishing, Jennifer Ahringer of Taylor & Francis, who I am looking forward to working with on many more projects—truly a pleasure. Thank you all.

The Author

Michael Holloway has 25 years of industrial experience. His background includes organic and polymer synthesis as well as laboratory to pilot-scale material manufacturing for Olin Chemical (New Haven, Connecticut); military and aerospace product development of sealants and coatings for Parker-Hannifin (Woburn, Massachusetts); product engineering microelectronic photoresists for Rohm & Haas Electronic Chemicals (Midland, Michigan); technical marketing and application engineering for General Electric Plastics (Waterford, New York); sales and marketing management of automated dispensing systems for Graco (Minneapolis, Minnesota); and most currently as director of reliability and technical development of industrial and heavy equipment lubrication for NCH Corporation (Irving, Texas). Through the course of his career, Holloway has been involved in training customers and associates, giving seminars, as well as authoring numerous papers and articles and presenting novel concepts and case studies at national conferences. He has served as a contributing writer for Manufacturing.net, *ASSEMBLY* magazine, *PlantServices* magazine, *Rendering* magazine, and *Lubrication and Fluid Power* magazine. He holds a BA in philosophy and a BS in chemistry from Salve Regina University, Newport, Rhode Island, and an MS in polymer engineering from the University of Massachusetts. Holloway has been master black belt trained in Six Sigma, served as an adjunct professor at the University of North Texas, and was managing editor for Porsche Club of America's *Slipstream*.

Chapter 1

Buy Bye

We buy for many reasons. If given the opportunity and authority to choose, we can make good decisions provided we take a little time and think it through. Some of the basic reasons why we buy are:

- To reduce the *time* and *labor* costs to fix, build, or make something more reliable
- To *save* money, both direct costs and indirect costs, by being more productive
- To improve *safety* and health
- For *cosmetic or design* reasons
- Because of the *prestige* of the product
- Because the product or service provides *value-added* contributions—get more than expected
- Because of being told to; because of *external influences* (boss, corporate policy, or warranty)
- For personal or *internal reasons* (experience or emotion)

You will be hard-pressed to find an example of something you have bought personally or professionally that does not fall into one or more of these categories. We will spend money, but we want to get something in return—a good product that does what it says it does or, better yet, it does better than the claim and you get some unexpected bonuses. The sad truth is that in many industrial settings, logic and reason are replaced by external influences or emotions. We would never purchase the cheapest house paint or engine oil for our personal use. If we did, we surely would not repeat the practice. That is not the case in maintenance, repair, and operations (MRO) spending. It is the industry norm to use price or brand names as a mitigating factor in the decision process. This is unfortunate, because price is generally a poor reason to purchase an item. Price is an easy metric to understand. Often the buyer will confuse price with cost. Worse yet, the cost to purchase and the cost to own are very different, indeed.

These buying decisions should be based on a combination of experiences and logic, but that is not always the case. There are instances when we buy specific items because we have no choice in the matter. The part may be specific to a tool or machine where the original equipment manufacturer (OEM) specifies the performance or dimensional aspects. Often there is little room for deviation. Sometimes, the decisions are taken out of our hands entirely. The corporate headquarters may have negotiated a deal with a national vendor for the lowest price, disregarding quality, performance, or worse, specifications. A growing trend in industry is to centralize the purchasing of maintenance, repair, and operation items. The corporate headquarters may negotiate the best price and delivery terms and pass that "savings" onto other divisions and departments. In theory, if the purchasing requirements are comprehensive, this can make sense if all products are of equal performance and the vendors are required to provide services that complement the mission of the organization. This book will explain how purchasing, maintenance, or the engineering departments can develop purchasing specifications that will ultimately drive reliability and reduce overall spending.

The common factor that influences a purchasing decision is typically price. The lowest bid gets the job. We all know that this is a very dangerous practice. Unless the engineering, production, and maintenance teams get involved in the purchasing requirements and part and product procurement specifications, this practice will continue; productivity along with reliability will never reach levels we all know are obtainable. What happens when other teams get involved and attempt to assist in the procurement process? The input of other teams can be invaluable.

The job of the buyer or purchasing agent has evolved over the past decade due to the Internet. More businesses are using lean manufacturing practices, which require just-in-time (JIT) inventories. If a manufacturing line is down because of a failed part or tool, then the company could lose tens of thousands of dollars, as well as customers. Some buyers or purchasing agents will attend conferences or trade shows for the products that take a certain degree of sophistication to purchase. These products are vital to the organization (such as an integral ingredient or a device). Many buyers rely on the requisitioner to know what they need for the nonvital items. Often a requisition will come in for an MRO item, and the buyer will resort to using a distribution house to provide the appropriate item. The decision comes down to price and availability. Many companies are now using procurement cards with success. Establishing this practice has streamlined the purchasing process. Most of the purchasing transactions are now automated, using electronic purchasing systems that link the supplier and firms together through the Internet. There is still room for improvement in this process.

The buyer has never been in a better position to demand performance from a vendor. It is important for buyers to require vendors to comply with their needs by making procurement specifications that not only take price and performance into account, but also contribute to the reduction of direct and indirect costs. The question remains, "How do I know just what to buy?" and more importantly, "How can I assure that what I am buying will actually help me?" The process is as follows:

1. Identify the reason for the purchase (relieve failure)
2. Identify what products are costing the most due to failure
3. Determine the performance needed to last longer
4. Choose what product test results and standards best suit your situation
5. Build the specification
6. Review product submissions
7. Select the product and monitor

For those who like flowcharts, consider the flowing process map (Figure 1.1).

Those engaged in the purchasing process have to have a sense for value and can appreciate that in certain instances a more expensive product will have to be purchased. This is because they recognize that the buying criteria are performance with

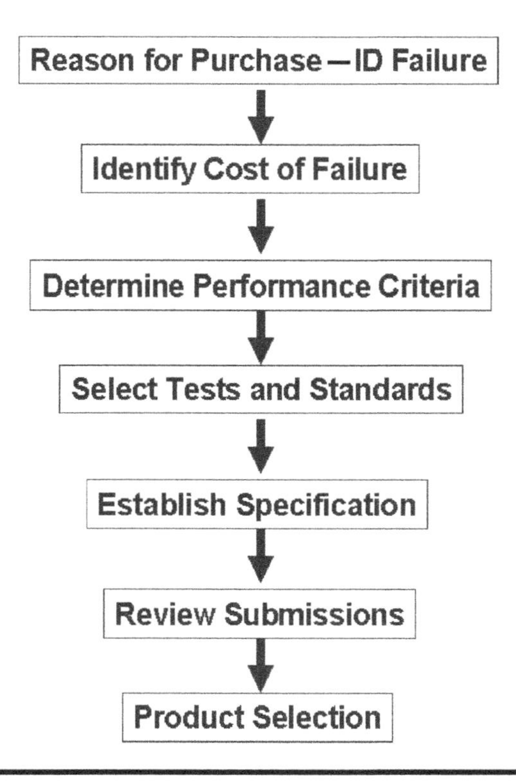

Figure 1.1 Holloway's purchasing specification development process.

the ultimate goal to drive down the cost of failure due to equipment downtime, repair labor, and parts replacement.

Each section of this book will build upon itself. This was done because you must first understand cost and value before you can appreciate the cost of failure. In order to properly document your product requirements, you must also understand why things fail. Understanding the fundamental sources of failure will help you to discern from the marketing collateral as well as the spin, hype, and misinformation that will be presented to you by vendors and associates. Yes, even your fellow employees will have a favorite product that they like, and it may be for all the wrong reasons. Having a firm understanding of cost, value, and failure will help you in countless ways. Applying your knowledge of value, cost, and failure will allow you to begin the process of establishing a specification.

Chapter 2

Buying to Save—Cost and Value

In this chapter, a method by which to calculate and determine the return on investment of a purchased item will be explained. To understand the concept, several topics will have to be understood. The goal is for readers to:

1. Understand performance and quality
2. Understand original equipment manufacturer (OEM) concerns such as warranties, recalls, and false claims
3. Define *operation* and *time costs*
4. Understand the basis of profit return from a purchase
5. Define *cost* and *value*
6. Understand the cost of failure
7. Understand how to calculate return on investment (ROI)

When a purchase is made, it is done with a specific intent—to replace an item or to improve upon a process with a new technology. The process can be complex, such as an integrated manufacturing line complete with vision sensors and eight-axis articulated robotic arms that position, weld, paint, and move large parts, or more simple such as repair of a pipe valve requiring a stainless steel bolt.

Several years ago, I surveyed several hundred maintenance staff members on why they bought various maintenance, repair, and operations (MRO) items.

The following summarize the reasons based on the points of view of technicians, mechanics, and a maintenance manager:

- To reduce the time and labor costs to fix something, build, or make something more reliable.
- To save money by being more productive.
- To improve safety and health.
- For design reasons.
- Because of the prestige of the product—I want the best.
- Product or service provides value-added contributions—we get more extras for the same price.
- We are told to; we have external influences (corporate policy or warranty).
- We buy for personal or internal reasons (experience or boss).

When I surveyed several hundred purchasing agents of the same companies, the answers were slightly different. What they believed to be the logical reasons for requisitioning or recommending a tool, material, service, or specific supplier or vendor were as follows:

- The requisitioned item is designed explicitly for a particular piece of equipment or job.
- It's the cheapest.
- This is what has always been done.
- The OEM has designated the replacement part or material, and we don't want to deviate.
- The requested item meets a performance specification called out by the OEM.
- The requested product or service is essential in maintaining continuity.
- The vendor has serviced our company for several years.
- We have a corporate agreement in place to only use this particular distributor for our MRO items.
- The suppliers we use for MRO understand our business; they have experience in similar work environments.
- That is who the maintenance manager told us to use.
- This is who corporate tells us to use. They have a national deal.

This initial survey yielded many responses, yet I was not able to easily get to the fundamental reasons why certain buying decisions were made. The survey was too open-ended. I needed to categorize and simplify the question of "Why do you buy what you buy?" If I could understand the prevailing reasons as to why, then I could adjust an inefficient buying pattern that would lead to greater cost savings. What I found out was surprising and reassuring.

A subsequent survey identified the reasons why certain product services were chosen over others. Three hundred twenty-seven people of various disciplines ranging from technicians and operators to chief executive officers (CEOs), working in manufacturing, mining, agriculture, and food-processing companies, were asked to force rank the reasons why they purchase items for work and personal use. What were the reasons for their selections? The survey was anonymous, carried out during various industrial trade shows, training seminars, and customer visits. The instructions were to force rank the reasons for any given purchase decision if left up to them. There could be no ties. A number had to be placed by each criteria, with one being the greatest reason why, followed by two, and so forth, with 10 being the least important criteria. A very surprising fact was that 86% of the participants answered the first six criteria the same. The survey revealed that performance and quality were the main factors. My job of establishing a way to develop purchasing specifications was going to be easy. The majority of respondents were already on board with buying based on performance and quality; they just did not know how to define it. The results of the survey are presented in Table 2.1.

This survey contradicts preconceived notions of purchasing practices. Typically, if you were to ask a manager why he or she thought items were bought beyond

Table 2.1 Reason for Purchase

Rank	*Criteria*
1	Performance: Product exceeds the required criteria compared to other offerings
2	Quality: Product performs consistently according to a required level of performance
3	Sales representative or distributor relationship
4	Vendor reputation
5	Technical support
6	Lowest price
7	Ease of purchasing: Online catalog available
8	Added services: Analysis, dashboards, seminars, customized solutions
9	Flexible billing terms
10	Delivery: Same day or next day

need, the manager would say it was the best price. Interestingly, many managers stated they would never buy an item purely on price but considered price as "the other guy's reason." I do not think I could find anyone who would purchase the least expensive house paint, and I do not think I could find anyone who would buy the least expensive tires for his or her daughter's car. In fact, with the exception of fuel and bottled water, I am certain that given the opportunity, most purchases are predicated upon experience and, in some cases, impression. That is not to say that discounted items and generic goods do not sell, they do. People will buy a product that is better in terms of performance and more consistent (higher quality) in terms of performance than other items. Companies with high customer satisfaction are not usually low-cost providers. In fact, many high-end providers such as Mercedes-Benz have repeat customers with a very high level of loyalty. You normally do get what you pay for.

Often, purchasing decisions that we make for our own homes or automobiles are based on experience, quality, and performance. It should be noted that there is a difference between quality and performance. As stated in the survey rankings, quality is product consistency according to a required level of performance. The product does what it is required to do for a long duration and is more reliable than any other product. A performance product exceeds a level of expectation deemed as status quo.

Interestingly enough, high performance may not mean high quality. Just because a product can perform better than other products does not mean it can do so consistently. Take, for instance, a top-fuel dragster. The engine that powers the racecar is the most powerful, producing speeds in excess of 300 mph and horsepower above 1000 hp. The top-fuel dragster engine is as high performance as one can engineer, but, unfortunately, the engine is typically good for only one race. It is common practice to rebuild a top-fuel dragster engine after each race. Performance is high, yet repeat performance is low.

Quality (or high quality) is a consistent level of performance that one grows accustomed to and eventually depends upon. Arguably a good example of quality dining would be the popular fast-food restaurant McDonald's®. The food served has a consistent preparation, allowing for very little in the way of flavor variation. You can have a meal in New York City one week and have nearly the same dining experience in Buenos Aires the following week and again the same dining experience in Beijing the following month. The consistency factor is the reason for the success of McDonald's restaurants. The customer (in this case, the diner) has established a certain expected level of performance for a fast-food meal. Any deviation would result in a letdown. Many scoff at the notion of fast-food restaurants being quality dining, yet the large percentage of fast-food restaurant customers are repeat customers from a very wide socioeconomic demographic. Price does not seem to be a factor. Actually, preparing one's own meals versus fast-food dining is roughly eight times less expensive.

A quality product shows very little deviation in properties but is susceptible to variations. Understanding variation influences will improve requested requirements. Raw material and human error aside, it is possible to understand how the manufacturing process influences the product's properties. A common index known as the *process capability index* (Cp) is often used. The formula for Cp is as follows:

$$Cp = \frac{\text{upper spec} - \text{lower spec}}{6 \text{ standard deviations}} \tag{2.1}$$

Another factor to consider is the influence on the testing accuracy of the product. The American Society for Testing and Materials (ASTM) test methods have gone through extensive statistical analysis to ensure that a certain level of repeatability and reproducibility is obtained. In certain instances, various industries have their own tests and methods to assess the performance of a product. These methods may not be as accurate as they should be. By understanding the test capability index (TCI), one can understand the accuracy of the testing methods of the product. The TCI is defined as follows:

$$TCI = \frac{\text{upper spec} - \text{lower spec}}{6 \text{ standard deviations of testing method}} \tag{2.2}$$

It is important to note that at least 25 data points are required to make the data statistically significant. The data provide information on how many analytical variations occur. Performance-centered procurement specifications are developed from technician and operations input. The performance requirements are determined by actively examining the various failures as they relate to the product in question and applying standards for the desired level of reliability. Typically, MRO items are considered a commodity product. Purchasing usually buys the cheapest item. It is only when maintenance, operations, engineering, and purchasing staff actually examined and understood the failure modes of various items and considered the total costs, did they realize that performance and quality (there is a difference) were actually driven by the item they were using. It sounds rather elementary, but in fact, it is actually a paradigm shift in thinking for many organizations.

The United States has always been a leader in innovation and inventions. The entrepreneurial spirit and creativity run throughout our corporate DNA. Companies large and small are always looking to produce the next great product to modify a production line for greater throughput, to reduce operating costs, or to produce a better-performing product. Sometimes, innovation can come from many different sources, and often it is completely unexpected. An MRO item can be a fastener, an adhesive, gear oil or a bearing, a portable screwdriver, or a welding rod. Every day the multibillion dollar market of maintenance, repair, and operations items introduces requirements that embrace innovation.

Innovation is defined as changes to a process or product that can be incremental (small or subtle changes to an existing material or method) or radical (paradigm shifts in technology or outlook), all with the eventual goal of solving a problem. Examples of technical innovation include General Motors' (GM) Lambda Vehicle Transmission (an all-wheel-drive system that utilizes unique software to optimize traction and handling). GM is one of the largest companies in the world, and millions of dollars and hundreds of engineers have been involved in the Lambda project. This technology will usher in a new era for GM transmissions. Another example is Z Corporation, leaders in the field of three-dimensional product duplication for rapid product prototyping. Z Corporation's products are like something out of Star Trek. Part dimensions are sent to the device, and in a short time, a three-dimensional plastic part is produced. This technology would have been science fiction a few years ago, and it is leading the way to further innovation. The foregoing are only two examples of hundreds if not thousands of examples that are occurring on a daily basis. Innovation can be found in very small, nontechnical companies as well.

As another example, a man I know (Dave) has been making homemade fishing tackle for years. He was having trouble finding particular worm and frog configurations and the colors to meet his needs, so he decided to make his own. Word quickly spread at fishing competitions, and soon he was making bait more often than he was fishing. A great source for bait ideas comes from many sources and often the answer is surprising and amusing. When I was very young, I remember watching an old-timer fly-fish. He would use a technique called "match the hatch," rooting around the grass, leaves, and even under rocks in order to identify which bugs were active and at what stage of development they were in. This expert would then tie a fly right at the shore by hand. This emulated the particular insect that was common for that pond or stream at that time of the year. This old codger would get a hit pretty much on every cast. Recently, Dave was walking around a pond by my place. The pond has plenty of catfish, but they can actually be rather picky, which is pretty strange for a catfish considering the average catfish will eat just about anything. He noticed a young boy by the edge of the pond with his fishing rod who was catching catfish like a pro. I knew the lad, who goes by the name Roscoe. He is the type of kid who is more comfortable outside than in, skinning his knees and catching turtles in the swamp. He goes through a pair of jeans a week and a pair of shoes each month. Roscoe's mother had packed him a bag lunch, which he was now using as bait. The catfish were going crazy over this stuff to the point of a feeding frenzy. Dave had to find out what it was.

Innovation can come from many directions. Sometimes it is a tweak of an existing product, other times it is a completely new idea that shakes the foundation of reason and makes the competition wonder, "How the heck did they figure that one out?" and still there are other forms of innovation that are nothing more than applied observations. Roscoe had taken time to observe and then apply his observations to solve a problem. The catfish were not biting his traditional bait, so he decided to observe what they would bite. He liked the cold pizza his mother packed for his lunch, and he thought the fish might like it as well. The fish liked it very much.

Roscoe really likes to fish, but he really does not like to wait around for a fish to bite. The little guy decided that the fish were not biting because they did not like what they were eating. He likes pizza, so he thought the fish might as well, and they did. He told me that he has tried other foods for bait, such as banana bread, apples, peanut-butter "samichs," and even chocolate kisses—all foods that he liked. The fish seemed to like the pizza the best, though. I think I know why. I did some quick product development and came up with some bait that used the essence of pizza and could be used repeatedly—the trick is the ... nope, this one I am not going to share (though I think Dave may have figured it out).

It is always surprising to see where the next great idea will come from. Roscoe had a need and an open mind. He observed his environment and tried something different that seemed to make sense to him, and it worked. That is innovation. Considering maintenance and repair, innovation in the way of tools and materials can make the difference between a company remaining profitable or having to close its doors. With the right tools and techniques, a job that may have taken several technicians a few shifts to accomplish may now take only one technician a few hours. When you consider the savings in equipment downtime and labor costs, innovation in tool and material development is as important (if not more so) as lean manufacturing or even a Six Sigma quality effort.

Concerning Warranties

One of the prevailing concerns with almost everyone I spoke with was the OEM recommendations and warranty concerns. This warrants further investigation as to the nature and the law of warranties. A common concern with various MRO components or materials is "voiding a warranty" because a specified material or component was not used. Below is an excerpt from the Magnuson–Moss Warranty Act. Often, an MRO product supplier will use the warranty as a reason to purchase a component. On the other hand, some suppliers use the Magnuson–Moss Warranty Act to try to educate the customer on "tie-in provisions." The following is an explanation of the Magnuson–Moss Warranty Act from the Federal Trade Commission (FTC).

On January 4, 1975, President Gerald Ford signed into law the Magnuson–Moss Warranty Act, Title 1, 101-112, 15 U.S.C. 2301 et seq. This Act, effective

July 4, 1975, is designed to "improve the adequacy of information available to consumers, prevent deception, and improve competition in the marketing of consumer products." The Magnuson–Moss Warranty Act applies only to consumer products, which are defined as "any tangible personal property which is distributed in commerce and which is normally used for personal, family, or household purposes (including any such property intended to be attached to or installed in any real property without regard to whether it is so attached or installed)." Under Section 103 of the Act, if a warrantor sells a consumer product costing more than $15 under written warranty, the writing must state the warranty in readily understandable language as determined by standards set forth by the Federal Trade Commission. There is, however, no requirement that a warranty be given or that any product be warranted for any length of time. Thus, the Act only requires that when there is a written warranty, the warrantor clearly disclose the nature of his warranty obligation prior to the sale of the product. The consumer may then compare warranty protection, thus shopping for the "best buy." To further protect the consumer from deception, the Act requires that any written warranty must be labeled as either a "full" or a "limited" warranty. Only warranties that meet the standards of the Act may be labeled as "full."

One of the most important provisions of the Act prohibits a warrantor from disclaiming or modifying any implied warranty whenever any written warranty is given or service contract is entered into. Implied warranties may, however, be limited in duration if the limitation is reasonable, conscionable, and set forth in clear and unmistakable language prominently displayed on the face of the warranty. A consumer damaged by breach of warranty, or noncompliance with the act, may sue in either state or federal district court. Access to federal court, however, is severely limited by the Act's provision that no claim may be brought in federal court if (a) the amount in controversy of any individual claim is less than $25,000; (b) the amount in controversy is less than the sum or value of $50,000 computed on the basis of all claims in the suit; or (c) a class action is brought, and the number of named plaintiffs is less than 100. In light of these requirements, it is likely that most suits will be brought in state court. If the consumer prevails, the consumer is awarded costs and attorneys' fees. Nothing in the Act invalidates any right or remedy available under state law, and most suits should proceed on claims based on both the Code and the Act.

Warranties that specify a particular brand name product for validity purposes must also provide that product at *no charge*. Obviously, most OEMs do not provide maintenance products for free. The most common way for a warranty to be valid is for the OEM to state that the product must meet industry-accepted minimum specifications. If a product meets or exceeds these specifications, then the warranty is valid. Therefore, warranties are not voided simply by the use of a particular brand, but rather because the material used does not meet a required set of standards or a specification called out by the OEM. OEMs with a major manufacturer listed in a service manual are not endorsing such brands, but simply have chosen them because they are well known and meet their minimum required specifications. In many instances, the service manuals are written by freelance technical writers.

If a replacement seal, O-ring, or gasket is required, the technical writer often will specify the material and as an extra include the manufacturer of the material. The technical writer will recommend the manufacturer, because it was easy for the writer to find a source. With that in mind, they believe it would take minimal effort for the customer to source the item as well. This practice is typical with lubricants such as grease, gear oil, and hydraulic fluids, as well as engine oils. In some instances, the OEM will have various items private labeled for them with a nonwritten, spoken requirement to use only their item or a warranty claim will not be honored. This is actually a common practice with many companies that produce capital equipment such as air compressors, articulated robot arms, or even engines. The equipment manufacturer may only sell a few of these items to a company. The sales and distribution force has to constantly find new business. Offering consumable items and service contracts allows for a revenue stream after the capital equipment has been received. The consumable items offered, such as filters, special fluids, or belts, are produced by a toll manufacturer and labeled with the OEM's brand. In many (but not all) cases, the item is no different than other items offered at a fraction of the cost by the manufacturer under their brand. An interesting aspect of the Magnuson–Moss Warranty Act is that it does not apply to warranties on products sold for resale or for commercial purposes. Does the act take into account MRO items? The question remains for legal interpretation.

The 1990s heralded in e-procurement, allowing many companies to streamline the purchasing process for maintenance and repair items. Along with this process, many companies also began to generate data, which allowed for spending to be analyzed. Based on these analyses, purchasing departments were now able to centralize the buying process at a corporate level, which then allowed them to buy across for multiple locations. The natural progression for such practices leads to a commodity-management approach to buying. This reduced the supply chain base and brought about national agreements. This series of events began to put purchasing on a slippery slope. Prior to centralized purchasing, many of the buying decisions that rested on the plant's purchasing agents' shoulders stemmed from direct involvement with the maintenance and production staff. Local suppliers, who had a stake in the business, were providing value beyond the mere adequate supply of a part or tool. By eliminating the local supplier, a valuable contribution to the business was lost.

National suppliers often have a distribution network that is more concerned with turning over inventories than providing value to the customer. There are many examples of these sorts of national agreements costing the company more in the long run through indirect costs than the savings the company had originally thought it would enjoy. To make matters even worse, in some cases, national suppliers would offer rebates to companies based on purchased volumes. These rebates would often be paid out to corporate officers. From time to time, a company's (or municipality's) upper manager would be indicted or convicted for accepting gifts exceeding an allowable level (many companies require acceptable gifts to be $25 or less in value). Such gifts come in the form of hunting or fishing trips, plasma TVs, and even cars. This practice provided incentives to continue the arrangement without any thought as to the actual cost of doing business with these national suppliers, or of the legal ramifications.

Although the process of automating and standardizing procurement through national accounts has the right intentions, greed and corruption can lead to individuals finding a way to profit. In a *Qui Tam* lawsuit first filed in 2001, consulting firms BearingPoint, Inc., Booz Allen Hamilton, Inc., Ernst & Young, LLP, and KPMG, LLP, settled (in January 2006) allegations that they had submitted false claims to various governmental agencies, including the U.S. Army, in connection with travel costs.

Qui tam is from the Latin phrase *Qui Tam pro domino rege quam pro si ipso in hac parte sequitur*, which means "Who sues on behalf of the King, as well as for Himself." Anyone aware of fraud committed against a government agency, a governmental department, or a branch of the military, can file a suit to recover the losses caused by the government fraud. Federal contracts require the contractor to give the United States the best price the contractor gives to any of its customers. The failure to do so is a form of fraud that violates the False Claims Act (see the Appendix for False Claims U.S. Code) and may give rise to a *Qui Tam* lawsuit. The False Claims Act can provide large incentives for whistleblowers that can get from 10% to 30% of the suit.

In the BearingPoint et al. lawsuit, each of the companies had received rebates on travel expenses that were charged to government contracts. The rebates were paid by credit card companies, airlines, hotels, rental car agencies, and other travel service providers. The investigation revealed that these companies had failed to fully disclose the existence of these travel rebates and did not reduce cost reimbursement claims by the amounts of the rebates on numerous government contracts. BearingPoint agreed to pay $15 million; Booz Allen agreed to pay $3,365,664; Ernst & Young agreed to pay $4,471,980; and KPMG agreed to pay $2,770,000.

False claims are taken very seriously by the government. For more information, please consult an attorney or your legal department. There are many people, companies, and organizations that defraud the government. Many attorneys estimate that over 10% of the annual budget actually is going to fraudulent institutions or individuals that are overcharging, submitting bills for work or items that the government did not receive, or basically overbilling for services.

JUSTICE RECOVERS RECORD $1.6 BILLION IN FRAUD PAYMENTS HIGHEST EVER FOR ONE-YEAR PERIOD

NOVEMBER 14, 2001, WWW.USDOJ.GOV

WASHINGTON, D.C.—The United States collected a record $1.6 billion in civil fraud recoveries during the past fiscal year, Assistant Attorney General Robert D. McCallum, Jr. of the Justice Department's Civil Division announced today.

"Taking a firm stand against those who defraud the government is always critical, and is especially so at this time," said Assistant Attorney General McCallum. "Taxpayer dollars should be spent on goods and services honestly provided, and benefit only those who are truly deserving and entitled. When they aren't, those dollars must be recovered. Our nation's security, health, and financial stability depend on it. This new record demonstrates the Department's continued commitment to pursue those who defraud the United States, whether by providing defective products, billing for phantom services, or otherwise misusing public funds for private gain."

Nearly $1.2 billion of the Department's settlements and judgments are related to cases filed under the federal whistleblower statute, which allows individuals who disclose fraud to share in the government's recovery. To date, whistleblowers have been awarded over $210 million for cases resolved in the past fiscal year (October 1, 2000–September 30, 2001). The government has strict guidelines and protocols concerning bid rigging, procurement, and contract awards. The private sector is not so fortunate. There are various laws designed to protect the buyer from price fixing and collusion as well as false advertising. The challenge is to be able to bring a suit against the accused and then prove your position and receive adequate compensation. It is a little-known fact that if a product a does not work correctly, the buyer not only has the right to have it fixed or replaced but, in some cases, has the right to sue for damages and attorney's fees. Unfortunately, the manufacturers have powerful lobbyists who work to keep many laws in favor of the supplier and not the purchaser. With performance procurement specifications in place, many performance issues can be avoided. Rarely will a supplier go through the work to comply with the specification, only to have it fall short and jeopardize future business.

Defining Operation and Time Costs

There are certain tasks that some folks do not enjoy doing—mowing the lawn, washing the car, or even painting the house. Many will do these jobs because they do not want to pay someone else to do them, or because they want to have greater control over the performance and quality of the job. Some people pay others to do tasks because it is actually more cost-effective. The time it may take to do a job could be better spent earning more than doing the self-appointed task would save. Still others will hire work out because they may want to spend their time doing something they enjoy and are willing to pay for others to do the work. Benjamin Franklin wrote that time is money. This concept is evident in any manufacturing plant, service business, and government agency, as well as in our personal lives. If one were to take a look at household activities, it would become obvious that even the most menial tasks could be streamlined to save time and frustration.

One task that is simple, yet provides frustration, is emptying utensils from the dishwasher. Granted, the dishwasher has saved countless hours when compared to the task of hand washing dishes, pots, pans, and utensils. We have all grown accustomed to the magic of automation. Many dishwashers even have built-in garbage disposals so one does not even have to rinse. Still, the task of taking the forks, knives, and spoons out of the bin and sorting them appropriately into their proper places in the utensil drawer is a very small source of frustration. A simple solution is to place all the forks in one wash bin, spoons in another, and knives in a different one. The time it takes to fill each washing bin is the same, yet the time it takes to empty the clean bin and put the contents away in their proper places is considerably faster and almost effortless. Using this time-saving method now makes putting away clean utensils simple. The practice of proper utensil placement was taught to me by my twin brother. He has become a master of streamlining tasks, driven by the desire to maximize his leisure time or increase his personal wealth by getting the most out of every minute spent working (professionally or domestically). He developed countless time-saving techniques that also reduce his wife's frustration (e.g., the practice of always putting the cover of the commode down after each use instead of leaving the seat up has, by his estimates, extended his marriage by at least 14 years).

First-year economics students are taught Bastiat's parable of the broken window. Frédéric Bastiat wrote the parable in his famous 1850 essay *Ce qu'on voit et ce qu'on ne voit pas* (*That Which Is Seen and That Which Is Unseen*). The parable is a good illustration for understanding the concept of hidden costs, or opportunity costs. The parable has fueled arguments in favor and opposition to the concept. Bastiat's original parable was as follows:

> Have you ever witnessed the anger of the good shopkeeper, James Goodfellow, when his careless son happened to break a square of glass? If you have been present at such a scene, you will most assuredly bear witness to the fact that every one of the spectators, that were there, even thirty of them, by common consent apparently, offered the unfortunate owner this invariable consolation—"It is an ill wind that blows nobody good. Everybody must live, and what would become of the glaziers if panes of glass were never broken?"
>
> Now, this form of condolence contains an entire theory, which it will be well to show up in this simple example case, seeing that it is

> precisely the same as that which, unhappily, regulates the greater part of our economical institutions.
>
> Suppose it cost six francs to repair the damage, and you say that the accident brings six francs to the glazier's trade—that it encourages that trade to the amount of six francs—I grant it; I have not a word to say against it; you reason justly. The glazier comes, performs his task, receives his six francs, rubs his hands, and, in his heart, blesses the careless child. All this is that which is seen.
>
> But if, on the other hand, you come to the conclusion, as is too often the case, that it is a good thing to break windows, that it causes money to circulate, and that the encouragement of industry in general will be the result of it, you will oblige me to call out, "Stop there! Your theory is confined to that which is seen; it takes no account of that which is not seen."
>
> It is not seen that, as our shopkeeper has spent six francs upon one thing, he cannot spend them upon another. It is not seen that if he had not had a window to replace, he would, perhaps, have replaced his old shoes, or added another book to his library. In short, he would have employed his six francs in some way, which this accident has prevented.

This tale tells of a shopkeeper's son who has accidentally has provided income for the window repairman (the glazier). The glazier benefits from the accident, and in turn, so does the baker who sells the glazier bread and the cobbler who repairs the glazier's shoes. The shopkeeper's son has kept the wheels of commerce turning, or has he? An argument can be made that there are hidden costs that the shopkeeper must absorb. Repairing the window may have resulted in the store being closed for a period of time. A closed store does not provide income for the shopkeeper. Additionally, spending money on a new window may help the glazier and others whom the glazier may employ, but the shopkeeper has to buy a new window with money he could have used to purchase inventory. With fewer items to offer his customers, a competing shop owner may benefit. The shopkeeper still has direct costs that have to be addressed, such as utilities and rent. The indirect costs that came from the accident are actually greater than the mere cost of a window.

Economics professor Ian Walker, of central England's Warwick University, developed a simple equation that provides an understanding of the Time-Equals-Money concept (*Time Is Money*, CNN.com). The equation can help someone understand how much his or her time is worth in relation to any task that person has to perform. From tasks at work to household chores, Equation (2.3) helps put time spent into a value:

$$V = \frac{W\left(\frac{100-t}{100}\right)}{C} \tag{2.3}$$

where V is the value of an hour, W is a person's hourly wage, t is the tax rate, and C is the local cost of living. Walker thinks that this equation will allow people to work out whether they are getting a fair rate for overtime as well as help them decide whether it is worth spending extra cash on various services to save time. When applying this concept, the tax rate t must be determined. Table 2.2 presents information obtained from the Internal Revenue Service (IRS) for a married couple filing jointly in 2007.

There are many reasons why we live where we do. It could be for employment or family. It may even have to do with personal interests such as leisure activities or a favorite sports team. Given the choice based on the pure value of your dollar, certain cities would be more appealing than others. Applying Walker's Time-Equals-Money equation, it is obvious that the value of an hour spent is considerably more in Memphis, Tennessee, than in Manhattan, New York. Table 2.3 establishes the 2007 value of an hour for a married person who earns $100,000 annually and files a joint tax return in the United States.

Of course, this equation does not take into account the possibility that your hourly wage may be larger than you truly deserve or much smaller than your potential. It is merely a stop-in-time indicator of the value of your compensation. Depending upon your salary, the equation may actually prove to provide a first-class justification to have someone else mow your lawn or to eat out more.

When it comes to MRO items, time is truly important in relation to how quickly a job can be completed and also how well the work gets done. Various tools can shorten the task, such as power hand tools or quick-drying adhesives. The supplier of such tools is responsible for proper training in the use of the devices and materials. It is not the job of the maintenance department to try to figure out how a new tool is best used or how to prepare a surface for an adhesive. Performance procurement specifications should not only include performance standards but should also specify operational requirements and application conditions.

Automobile mechanics and many technicians supply their own tools. It is not uncommon for a maintenance technician in a manufacturing plant to own

Table 2.2 Income and Tax Rate

Income	*Tax Rate (t) (%)*
From $0 to $15,100	10
From $15,100 to $61,300	15
From $61,300 to $123,700	25
From $123,700 to $188,450	28
From $188,450 to $336,550	33
From $336,550 and above	35

Table 2.3 Cost of Living Index per State

City	*Cost of Living Index (C)*	*Value of an Hour (V)*
Memphis, Tennessee	89.9	$55.62
Topeka, Kansas	92	$54.35
Springfield, Illinois	92.1	$54.29
Jackson, Mississippi	92.3	$54.17
Omaha, Nebraska	92.3	$54.17
Des Moines, Iowa	94.4	$52.96
Cincinnati, Ohio	95.1	$52.57
Dallas, Texas	95.2	$52.52
Montgomery, Alabama	96.1	$52.03
Salt Lake City, Utah	97.3	$51.39
Billings, Montana	97.5	$51.28
Phoenix, Arizona	98.2	$50.92
Buffalo, New York	98.8	$50.61
St. Louis, Missouri	99.9	$50.05
Richmond, Virginia	101	$49.50
Milwaukee, Wisconsin	101	$49.50
Cleveland, Ohio	102.7	$48.68
Denver, Colorado	103.5	$48.31
Baton Rouge, Louisiana	104.1	$48.03
Cheyenne, Wyoming	107.7	$46.42
Eugene, Oregon	109.5	$45.66
Seattle, Washington	118.6	$42.16
Chicago, Illinois	128.6	$38.88
Juneau, Alaska	131.6	$37.99
Boston, Massachusetts	136.8	$36.55

(Continued)

Table 2.3 Cost of Living Index per State (Continued)

City	*Cost of Living Index (C)*	*Value of an Hour (V)*
San Diego, California	141	$35.46
Los Angeles, California	153.1	$32.66
Honolulu, Hawaii	162.4	$30.79
San Francisco, California	177	$28.25
New York (Manhattan), New York	212.1	$23.57

thousands of dollars worth of tools. When technicians spend, it is money out of their pocket, and they are very savvy as to which brand to buy and to what specification. There are some instances when the maintenance department will purchase a special tool for a particular machine (e.g., a 2000 ft/lb torque wrench) or a device that will be used for diagnostics, such as an infrared camera or a vibration analysis spectrograph. The technicians typically would not invest thousands of dollars of their cash into such a tool unless it was going to be used often enough to give back a return on their investment. This mentality spills over into management.

When a technician or maintenance manager reads about a new device, material, or tool, typically the first impression is not unlike a child viewing a new toy commercial. It is human nature for people who use their hands along with their minds to enjoy the latest gadget or feature. It is human nature. Marketers of MRO items take full advantage of this characteristic. The Nobel prize–winning poet Rudyard Kipling wrote: "I keep six faithful serving men who teach me well and true. Their names are What and Where and When and How and Why and Who."

In marketing, this Kipling pronouncement is called the "5W1H" formula. It is very effective and clearly gets the message across. The following questions are answered in marketing material and advertisements:

- *Who* are you trying to sell to?
- *What* does your product do for them?
- *Why* is it superior to other products?
- *How* can you prove this contention?
- *Where* should you advertise to reach the buyers?
- *When* is the best time to reach them?

Open a trade magazine and see how many top-notch companies use this formula to try to entice a buyer. You may also notice how many companies that produce and sell MRO items do not subscribe to this formula. Chances are that the companies using these forms of marketing are also the ones that have a greater share in the marketplace, but is this justified? Should companies that market better have better sales, even if their products may be inferior? Sadly, all too often, the best-marketed product is not necessarily the best product in terms of performance, quality, and return on investment. Once the buyer's interest is piqued, the seller provides enough marketing collateral, such as data sheets, copies of trade magazine articles, and even commercials, in order to bring a sale closer to closure.

Hype and spin can take a supplier pretty far; nevertheless, the vendor still needs to provide products that will hold up to the claims. Advertising is regulated by the authority of the Federal Trade Commission U.S. Code, chapter 15, section 45, which prohibits unfair and deceptive acts or practices. (See the Appendix for U.S. codes on False Claims and False Advertising.) It takes just a few mistakes on the product manufacturer's part to destroy many years' worth of goodwill and marketing if their product is not used according to the intended application.

Take, for instance, the following example. Powers Fasteners is considered the largest domestic manufacturer of quality concrete fasteners in the United States. Powers offers a wide range of mechanical, adhesive, powder-actuated, and gas-fastening systems. They also specialize in plastic and hollow wall anchors, carbide drill bits, and roofing fasteners. When Fred Powers, Sr., began his career with Rawl USA in 1945 as a sales manager, the company had one plant, two warehouses, and eight employees. In 1957 when Fred Powers, Sr., purchased Rawl USA, they had increased the staff to 19 and had 9 warehouses. In 1994, with 3 plants, 30 warehouses, and 285 employees, Rawl USA changed its name to Powers Fasteners. In 2006, Powers Fasteners had 4 plants, 36 warehouses, and over 400 employees. That same year, concrete ceiling panels on the Ted Williams Tunnel in Boston, Massachusetts, fell, killing a motorist. The epoxy fastening system provided by Powers Fasteners was suspected to have failed.

In the days, weeks, and months that followed, more of this investigation was played out through news organizations, in public forums, and on the Internet. It will not be until those determined to be responsible for this disaster pay the price that the door will be closed on this engineering mishap. One thing is for certain, with the national coverage this situation generated, laws will likely be changed, and the Big Dig tragedy will serve as an example of what not to do. Hopefully, this action will prevent another life from being lost due to substandard construction and design.

It is not fully understood who is to blame for the failure of the ceiling panels. Was it the materials? Was it the application method? Did the contractors rush the job and not use the correct amount? Was it a lack of a definitive performance specification, which not only would require physical standards to be met but also manufacturing quality standards as well as supplier value contributions such as training?

The Big Dig tunnel project cost over $15 billion—an amount much greater than the original bid. A life was lost, possibly due to a failed fastening system—an MRO item. The cost of the failed fastener epoxy system that Powers supplied was less than $2000. The courts will soon pass judgment.

There are times when a supplier recognizes the fact that their product was manufactured with flaws in design, materials, or workmanship, or was tampered with, which could or did result in lack of performance or safety of the user. When such an event occurs, the manufacturer will recall that model or batch (see the Appendix for the Recall Handbook). Recalls can cost companies millions of dollars. The recall can involve a specific batch, an entire production run, or worse, a model line. A product recall can be very costly, often requiring replacement of the entire stock of items affected or payment for any damages that may have occurred that were caused by the product. Worse yet is the aftermath of a recall, when any trust that had been established by the manufacturer is now shattered. In 1982, Extra Strength Tylenol capsules that had been laced with potassium cyanide poison resulted in the death of seven people. The manufacturer of Tylenol, Johnson & Johnson, lost a 35% market share of this painkiller, but fortunately, due to the company's quick and aggressive action, the company was able to rebound from the 8% market share outcome that was the result of the sabotage. The incident brought about the advent of tamper-resistant packaging.

Note: The Office of Compliance at the U.S. Consumer Product Safety Commission has prepared a product recall checklist to help manufacturers, importers, distributors, and retailers conduct an effective and comprehensive product safety recall. This list is a guide only to assist recalling firms in removing products from all stages of the distribution process. Not all items will necessarily apply in all recalls. Recalling firms may obtain further guidance from the Office of Compliance at 301-504-7913 or at sect15@cpsc.gov. Information can also be found at www.cpsc.gov.

Some Recent Maintenance, Repair, and Operations (MRO) Recall Items

On July 10, 2007, Milwaukee Power Tool recalled about 1 million batteries used with Milwaukee Power tools because they could explode.

> Included in the recall are Milwaukee Power Plus, Chicago Pneumatic, and Extractor 14.4- and 18-volt 2.4-Ah NiCad battery packs. If a vent on the battery cell is damaged or compromised during use, the battery can explode and pose a laceration hazard to consumers. Milwaukee Electric Tool Co. received 35 reports of incidents, including 11 injuries from battery packs exploding while in use. Injuries include minor cuts, bruises, and some hearing loss.
>
> The recalled batteries are designed to provide power for drills, saws, radios, flashlights, wrenches, and Extractor windshield glass removers. The recall includes 14.4- and 18-volt, 2.4-Ah NiCad Milwaukee Power Plus, Chicago Pneumatic, and Extractor battery

packs, manufactured between July 1999 and February 2004. The brand name can be found on a label on most battery packs. However, some 14.4-volt, 2.4-h packs did not have "Power Plus" on the label. The battery packs were sold both with tool kits and as individual battery packs. Battery packs manufactured after February 2004 are not included in this recall.

The battery packs were manufactured in Mexico and sold by home centers, hardware stores, industrial distributors, and vehicle service distributors nationwide, from July 1999 through 2005.

Consumers should immediately stop using the recalled battery packs. Consumers should contact the Milwaukee Electric Tool Co. to determine if they have one of the recalled batteries. The company will provide a free replacement battery pack for consumers with recalled units.

On the Road to a Profit Return on a Purchase

Maintenance managers have operating budgets. When a request comes in from the floor for a new tool, device, or material, it is up to the manager to determine whether it makes sense in terms of an ROI. Some basic ROI criteria to be addressed are does the tool, device, or material

- Reduce the time it currently takes to do the work?
- Reduce manpower/labor to perform the work?
- Provide a design improvement over current tool, device, and material?
- Provide a safety improvement over current tool, device, and material?
- Have increased longevity?

Return on the investment for a new cordless screwdriver or a bucket of grease may at first seem ludicrous, but in actuality, it is not. Consider the following excerpt from the author's unpublished, private novel entitled *American Standard*:

> DX met Doug Church on his second trip to Ames Forging. Ames has been in the metal forging business for over 140 years. They made their money supplying various tractor and implement manufacturers with plows and cultivating discs. They expanded their offering to include various armored components for the military. This business decision was made due to the cyclical nature of war and agriculture. Both seemed to play off each other, which afforded Ames consistent work.
>
> Doug had worked his way up through the production ranks. He began working at Ames during the summer of his junior year of high school and attended the local college nights in order to earn his degree in Engineering Technology. Doug was recently appointed Assistant

Plant Manager and was exceptionally eager to prove his worth to the Plant Manager, the General Manager or any suit that came in from the corporate headquarters. Doug's wife was expecting their second child and it was always the running joke on the production floor that the child was conceived while he was asleep because Doug spent all his waking hours at the plant. He worked hard, meant well, and only wanted to provide for his family as best he could.

Doug's superior was Tommy Kohler the Plant Manager. Koh (the name everyone knew him by) was a man with a rather abrasive personality and little patience for management. He was under the opinion that the sales force made outrageous demands and forced the company into producing product that was not their strong suit. Koh relied heavily on Tony Colletti who was in charge of maintenance. The two grew up together, played high school football together, were each other's best men, and were each other's first son's godfathers. The plant ran because of Koh; the machines ran because of Tony.

It was two months and four shifts into Doug's new appointment that Koh pulled him into his office. DX was on-site working with Tony on a reliability study for the hydraulic presses. Koh wanted Tony and me to be in the office when he addressed Doug. "Dougy, we got suits coming in. Rumor has it they are from some holding company looking over the joint and maybe even looking into purchasing it. Now I could care less about all that but it may give you a good opportunity to shine at Mahogany row up north." Koh referred to the corporate guys as Mahogany row and disdained them. He was a year or two from retirement and would take it sooner if given a reason. Ames had been going through restructuring and it was no surprise that the forging division might be spun-off. "I need you to focus on the five presses and make sure we have 100% capacity for the next few weeks." Koh looked at Doug as a son and had a stake in his success. Doug was married to Koh's niece.

"I'll give it my total attention. Thanks for the heads up sir," Doug answered.

"Tony and DX here will help you out. Let them know what is what, if any problems occur. I will be at the lake house for the next two weeks starting as soon as I get my coolers filled. Page me if you get into something thick. Just don't do anything stupid!"

Doug now had the opportunity he needed. Koh would most likely retire before the buyout and the new owners would prefer to keep the present manufacturing staff in place until they could assess if a change was needed. With hard work and some practical understanding, Doug could weather the storm and survive to become the next Plant Manager. Doug had many things going for him: experience, education, and personality. Doug also had a major fault that would prove to be disastrous;

he had certain opinions of things that were based on appearance and not function.

They left Koh's office; Doug immediately sprang forward and was five steps ahead of Tony and DX. As the two walked further behind Dougy, Tony shook his head and laughed. "Doug has wanted to shine for a long time. If he is smart, he will just let the presses run and not do anything. They are running at 90% now and the only reason they aren't at 100% is the operator's decision, not the machines." DX nodded but remained silent.

Tony looked at Doug as he pulled even further away. "Give a man as much rope as he needs to hang himself with and he will." He said laughing. "Let me know what you need me to do DX." DX just smiled.

DX had a strange and strong feeling that things were going to go south and there was nothing Tony or anyone for that matter could do but watch and wait. It was the lull before a storm. The stillness you feel right before a tornado sets down. The Ames sky was turning a peculiar shade of green with a muffled rumble in the distance.

DX met with Doug later that day. Doug had scheduled a meeting which was not necessary. DX was on site anyway and used a desk that was just outside of his office. "DX, I was thinking that during the shutdown this weekend we should have the presses power washed and then repainted. I have a PM coming up for an oil change out and a seal rebuild. This might give us a chance to nail down a few things all at once. What do you think?"

"Diamonds shine, presses press."

That was something Doug did not want to hear. "The presses look awful. These bankers like clean factories. Get me the amp, vibration, pressure, and temperature readings as well as the throughput numbers. If we ramp up over the next several shifts, we can hit our production quota and then some. This will give us a cushion in case we need an extra shift to complete the work."

"So said," DX said and left his office. At the end of the day, he got paid either way.

There are many reasons why things fail. All things will fail and there is always a reason. Doug was going to learn this firsthand. Call it paying tuition to the school of hard knocks. Tony happened to retell this story many months later while being interviewed by a trade magazine. Tony didn't consider this story out of the ordinary; after all, he made a career out of the folly of lesser men. Still, he couldn't help but wonder just what would have happened if people acted in accordance with sound engineering principles and not emotions. It was shortly after the last coat of paint had dried on the presses at Ames that it began to peel off. No one noticed at first because it was 3:00 A.M. The only one in that section of the plant at that hour was the security guard who was fast asleep at his desk.

Manufacturing resumed the following day. The presses could not be shut down again so the order was given to blast off the peeling paint with compressed air. Funny thing about paint chips though, they seem to get all over the place. To add insult to injury, another order was given to wash down all the equipment again. This washing proved to not only contribute to the flash rusting of the freshly painted/freshly peeled equipment, but the paint chips also provided an assorted array of calamity for the powder-coating department to deal with, as well as a fantastic way to clog the drains. Murphy had his hand in this one. Not only was there a drastic decrease in the production quota for the next week due to the drain issue, but the quality control department was having a field day with all the rejects from the coating line.

It was never determined if the metal was not properly primed or if the paint that was used was of poor quality. Weeks later there was speculation and finger-pointing and none of it provided any enlightment. Of course, this all had to happen when the potential investors were taking their tour. Needless to say, the investors were not impressed and decided to pass on the plant. Young Doug had some "splain'n" to do. The upside, Wal-Mart was hiring.

For many plant managers, the maintenance department struggles to communicate the benefits of various requisitions in terms that upper management can understand. Making a business case for an expensive device, material, or tool best have a tangible deliverable. This is critical if upper management includes accountants and financial analysts. It is important to understand that every MRO purchase decision has an impact on the company's profitability and future. It would not be a surprise to anyone in maintenance management that a major barrier to a company's success is the lack of upper management's understanding of the maintenance strategies. To make matters worse, many companies do not measure maintenance performance or even understand how to establish a method for calculating and reporting a return on an investment in a material, device, instrument, or tool.

Costs and Value

Money, get away. Get a good job with good pay and you're okay.
Money, it's a gas. Grab that cash with both hands and make a stash.
New car, caviar, four star daydream, think I'll buy me a football team.

("Money" by Roger Waters, Pink Floyd, from the album *The Dark Side of the Moon*, produced by Pink Floyd, released March 1973, recorded in Abbey Road Studios between June 1972 and January 1973)

The worth of something is what someone is willing to pay for it. The price of something is what the market psychology will tolerate, but the cost of something is the energy that it takes to produce a product or a service and make a profit. This energy is made up of three different cost types: *direct costs*, *indirect costs*, and *standard costs*. The three different types of costs can fall into three different classifications known as *cost behaviors*: *fixed costs, variable costs*, and *semivariable costs*. There are many examples of different costs associated with a product, and it is common to have the various direct costs fall into different cost behaviors. In the procurement purchasing function, it is important to understand where the cost basis originates. It is even more important to understand how a good or a bad purchase can influence the cost of a product, the profitability of the company, or the balanced budget of a municipality.

The cost that is easiest to track and influence is *direct cost*. The *direct cost* is defined as a cost that can be measured and allocated to a product or service. Material costs are often directly related to the production of a product, as is the labor involved in the assembly. The *indirect costs* cannot be allocated to a product or service, per se, rather they support the overall operation. Common tools, general supplies, and even equipment maintenance are considered indirect costs. Companies, the military, and even the government attempt to capture indirect costs and associate a percentage to a specific cost center. Even with the best cost accounting methods, this can be a difficult task. The third type of cost is *standard cost*. Establishing the actual cost of production or service prior to event is *standard cost*. The *standard cost* is developed from anticipated labor rates, material costs, and overhead costs according to a predetermined level of production. The *standard costs* play a very important role in cost containment and control. These costs are often used to prepare bids, estimate future manufacturing costs, and establish the value of WIP (work-in-process) of produced items.

Manufacturing Costs

Any time something is manufactured, material, machines, and people are required. These things cost money. The cost incurred combined with the profits account for the selling price. There are different elements to the manufacturing costs that have a direct influence on the profitability of a company. When a production line stops due to a bearing failure or a piece of mining equipment is held up in the shop for hydraulic repairs, the company is not making profits but is still incurring a variety of costs to operate. A small failure can drive a company's profitability down dramatically. If the failure is constantly repeated, it can be the contributing factor for the demise of the company. For improved reliability to occur, it is important to understand where various costs originate. Often an intelligent buying choice for a small item can have a very large impact on the company's profitability and future.

Manufacturing costs can be broken down into a few categories; *material costs* (*direct materials*), *labor costs* (*direct labor*), and *manufacturing overhead*. *Direct material costs* refer to any material (wood, steel, plastic, paper, glue, screws, paint, etc.) used to make a final product. The finished product from one company can be the raw material for another. Often there can be many degrees of separation from the raw material to the absolute final product.

Material Costs

Consider the laptop computer. A laptop computer is made up of the case, the electronic components, the power source, the screen, as well as the packaging and the user manual and software. The electronic components find their roots in sand, which the silicon chips are made from; various metals that are mined, refined, and further processed; and crude oil that is refined to make the chemicals used to make and etch the circuit profiles. Consider the fact that the various plastics used in the housing, circuit boards, and integrated circuits are derived from crude oil as well. The user manuals are printed on paper (from wood), and the ink is either a vegetable dye or derived from petrochemical feedstock also from crude oil. A closer examination would reveal that there are actually many tiers involved in laptop computer production as well as many different materials directly related to the end product (also known as *direct materials*).

Many years ago, a company would try to accomplish all the tasks required to produce a product—from clearing timber to producing its paper product, from mining iron ore and casting the metal to fabricating the product. Companies were completely vertically integrated. Over the past few decades, there has been a specialization in industry. There are companies that are just involved in one small aspect of the product. Take, for instance, the computer industry. Some companies just produce the packaging of the computer, and others engage in final assembly where the keyboard and screen are attached to the case. Prior to final assembly, there are companies that build and assemble the subcomponents, for instance, the memory chips and the various drives. Before the subcomponents are made, the various devices have to be fabricated, such as the integrated circuits and the disc reader. There are companies just involved in that function. Before the device fabrication takes place, various electronic parts must be processed, tested, and finished, and the silicon metal substrate has to be prepared for an electronic microchip to be built. This is another example of a specialized function. The silicon wafer used for chip manufacturing is made from a solid block of silicon metal that has to be refined. The silicon metal is made from sand that requires processing. The sand is dug from a quarry or is mined.

The idea that a laptop computer has hundreds of companies involved in its production is amazing. To think that the laptop has its roots in sand is almost impossible to imagine. Raw materials are mined, pumped out of the ground, or grown. In the truest sense, the only real production is mining, oil drilling, and farming. All other activities are merely services that manipulate raw materials.

Direct material costs can have a major influence on the cost of the final product (as in ruby earrings) or can be insignificant (as in a painting). There are thousands of steps involved in the production process of many items. Producing only a few of a specific item such as a laptop computer would cost millions of dollars. The laptop computer is affordable because millions are produced annually. The enormous volume keeps the price down. This is known as the economy of scale.

Labor Costs

Production requires labor. Even the most automated facility requires operators and technicians. *Direct labor* accounts for a percentage of the overall cost of the product. A labor-intensive product such as a Rolls Royce automobile is more expensive, whereas a can of a soft drink is relatively inexpensive. Soft drink bottling is almost completely automated. Still, the people involved directly with the production of the final product have to be compensated; therefore, there is a cost involved. It is a common practice for many companies to lay off workers in order to achieve higher profitability. Material costs have little if any flexibility, and other overhead costs are fixed or very inflexible. Labor is often the one element of the cost pool that can be acted upon. Unfortunately, this is routinely met with reduced company morale eventually leading to production and quality issues, not to mention customer resentment. There are countless examples of companies that artificially inflate profitability by reducing labor costs. This may be favored in the short term by shareholders and Wall Street, yet it always seems to backfire when the dust settles. General Electric and Ford are very large companies that have taken this approach with very questionable results. The energy and emotion applied to downsizing could have been better applied at the beginning by developing better products and setting up proper systems and standards to ensure that layoffs would not happen.

Manufacturing Overhead

The third category is *manufacturing overhead.* These costs include everything involved in the production of the product except the direct materials and the labor. It is generally a catchall that includes utilities, building depreciation, maintenance, indirect materials, indirect labor, taxes, and insurance, to name a few. Manufacturing overhead is very difficult to trace back to a specific product unless there are only one or two products produced, which is hardly ever the case. These costs are typically fixed costs with little or no fluctuation. Although these costs are predictable, in many cases, they are contracted to have a set expense. They will continue to incur even when there is a production shutdown or an employee strike. Another thing to consider is that certain overhead costs are independent of the amount of volume produced. For example, the depreciation of the factory or the rental cost of equipment or instrumentation is independent of production.

Certain materials used in the final production of the product but which are relatively small in terms of material cost and usage may be lumped into overhead costs and not raw material cost. These are called *indirect materials*. An example would be the sprinkles on an ice-cream cone. The cost is insignificant compared to the cost of the ice cream, the cone, the labor to make the ice-cream cone, and the rent for the ice-cream parlor, all of which account for a considerable portion of the product costs. The same can be said for indirect labor—various administration functions that may indirectly support the production effort but are not directly involved in the production. These tasks are *indirect labor* costs. Examples of these functions include supervisors, security guards, and administrative assistants. These functions are critical but are very difficult to trace to a specific product. In certain cases, indirect labor can become overinvolved in the costs of manufacturing. Later we will see an example of what happens when a seemingly small failure on the production floor quickly involves indirect labor and the ramifications that can lead to a catastrophic event.

Additional Costs

The other type of cost to consider is nonmanufacturing cost. These additional costs support the manufacturing operation: research and development, selling expenses, and administrative costs. The research and development role often overlaps into production. This cost is often combined into overall direct operational costs. The selling expenses associated with a company include all functions necessary to create the need (advertising and marketing). The administrative costs include accounting functions, customer relations and service (sales or distribution), technical assistance (technical service), product and service manuals (technical writing), sales support functions including the accounts payable and accounts receivable departments, the shipping and receiving function, and all the various levels of management and clerical help involved in these departments.

Cost Behavior Patterns

Accounting for where monies are spent and tracking the usage of labor and material allows organizations to make strategic choices. The more defined and exacting the accounting practice is, the greater the likelihood management can be provided with the tools needed to build the business. The two most common cost behavior patterns are *fixed costs* and *variable costs*. *Fixed costs* are also known as the *capacity costs*. These costs do not change within an agreed-upon time frame. Examples of these expenses would be rent or mortgage (provided the percentage rate is fixed and not variable), property tax, insurance, administration salaries, as well as various fees associated with running the business.

The direct labor and the material costs are examples of *variable costs*. These costs are associated closely with manufacturing output. As production of a product increases, the cost will also increase but so will profits. Additional costs can result from the increased production. Often when production is increased, so is the opportunity for equipment failure. These costs are known as *incremental costs* or *incremental revenue*. These costs or profits are very difficult to determine and normally involve a limited, short-lived change. For example, the incremental cost for driving a delivery truck may be $0.47/mile. This cost depends on such factors as the driving conditions, the driver's technique, traffic, the age of the engine, the quality of the fuel, as well as the condition of the tires. Many factors and variables influence the real cost. Examining the actual cost may require many months if not years worth of data and then performing sophisticated analysis to determine if the variables have a significant influence on the overall cost of operations.

While writing this chapter, our son Max became interested in what I was doing. Actually, I think he just did not want to go to bed, so he took an immediate interest in what I was doing and wanted it explained in terms he could understand. What 8-year-old boy does not understand ice-cream cones? I decided to use an example from Chan Park's *Contemporary Engineering Economics*. I explained to our son that everything costs a certain amount of money to make. You charge a little extra so you can make a profit to pay yourself for the work done and also to invest back into the business. Let's take a look at Chan Park's example of an ice-cream parlor (Table 2.4).

The unit price was figured out by having a target of producing 185,000 cones annually. This volume would generate $462,500 in sales. This production volume was needed considering the fixed costs associated with running the business as well as the material costs that fluctuated. To achieve this sales requirement, you would have to sell at least 500 cones a day. Max and I agreed that was a lot of ice-cream cones to sell. I asked him what we could do to increase our income if we could not sell that many. He thought we should raise the price to $3.50 a cone. This is a common technique, but I explained that it may not be the best approach. When I asked him if he would like to spend $3.50 for an ice-cream cone, he replied, "That better be a good ice-cream cone Dad!" I had to agree. For a $3.50 ice-cream cone, it better come with a song and a dance. This image brought about a flurry of examples of ice-cream cone songs and dances that did not get us any closer to the problem at hand or me finishing the chapter.

With all the various costs associated with running the business of ice-cream cones, perhaps a devastating thing to happen would be a power outage. Sometimes, the smallest of failures can have a large impact. Consider the example of the polyvinyl chloride (PVC) pipe joint that failed. This pipe fed the boilers softened, chemically treated water that then supplied steam to several steam turbines that, in turn, generated power to the city of Dallas, Texas. The steam turbines came to

Table 2.4 Ice-Cream Parlor Costs

Items	*Unit Cost*	*Percent (%) of Price*
Ice cream	$0.65	26
Cones	$0.05	2
Rent	$0.61	24
Wages	$0.25	10
Payroll taxes	$0.05	2
Sales taxes	$0.23	9
Business taxes	$0.08	3
Debt service	$0.23	9
Supplies	$0.09	4
Utilities	$0.08	3
Other expenses (insurance, advertising, fees, etc.)	$0.05	5
Profit	$0.13	5
Total	**$2.50**	**100**

a halt one by one. A $17 item cost Dallas inhabitants and business owners several millions of dollars in lost revenue. The largest contribution to the cost of the product would be ice cream at 26%. As far as we were concerned, a power outage would shut down the freezers, and the ice cream would melt. Recuperating the lost profits would take a while.

After spending some time explaining the various types of costs, we took to the task of assigning them to the costs in our ice-cream example. Needless to say, our son was fast asleep within 10 minutes.

Examples of "Wasted" Cost Overruns—The $600 Toilet Seat

One area that we are hypercritical of is the apparent wasteful spending of the government and the military. Often the media looks to expose spending corruption. Being the son of a man who spent over 30 years as a federal government civil servant for Naval Underwater Warfare, I am keenly aware of the challenges that our government and military face as well as some of the apparent shortcomings. One thing is very obvious when dealing with the government and the military, there are several layers of redundancy built into almost every device, activity, and process. If there is one thing our government understands, it is failure. In the military, there is absolutely no room for failure. If, by chance, something was to go wrong, the energy and time invested in the "fix" are enormous. An example of this is the fabled $600 toilet seat. In 2004, Senator Chuck Grassley (Republican, Iowa) said, "I exposed the spending scandal in the '80s when federal bureaucrats saw no problem in spending $600 for a toilet seat." Some now claim that neither that nor his also famous revelation of the Pentagon spending $400 for a hammer actually ever happened. Others say the prices paid were fair and justifiable. The $600 toilet seat was determined to be "fair and reasonable" by a Naval Contracting Officer, based on his detailed knowledge of the manufacturing processes and degree of effort known to be required from the vendor to manufacture this item.

The U.S. armed forces are often in the position of having to make equipment that lasts decades longer than the original design. For example, the B-52 bomber is more than 50 years old and expected to be useful for another 20 years. The famous toilet seat came about when about 20 Navy planes had to be rebuilt to extend their service life. The onboard toilets required a uniquely shaped fiberglass piece that had to satisfy specifications for the vibration resistance, weight, and durability. The molds had to be specially made as it had been decades since the original production of the planes. The price of the "seats" reflected the design work and the cost of the equipment to manufacture them. The problem arose because the top-level drawing for the toilet assembly referred to the part being purchased as a "toilet seat" instead of its proper nomenclature of "shroud." The Navy had made a conscious decision at the time not to pay the OEM of the aircraft the thousands of dollars it would take to update their top-level drawing in order to fix this mistake in nomenclature. Later, some unknown Senate staffer combing lists of military purchases for the Golden Fleece Awards found "Toilet Seat—$600," and trumpeted it to the news media as an example of "government waste." The Senate then wrote into the appropriations bill that this item would not be purchased for anything more than $140. The shroud has never been purchased since, as no one can make the shroud at that price. President Reagan had actually held a televised news conference during which he held up one of these shrouds. During the press conference, he explained the true story. The media of the time, and still today, incorrectly reports that the Pentagon was paying $640 for a $12 toilet seat.

The government pays certain jobs on a cost-plus basis—the cost of the item plus a reasonable markup. The reason the prices are so high is the manufacturer is being allowed to allocate its depreciation expense for its equipment equally to each item the manufacturer sells. Therefore, a wrench "costs" $500 to make. The rationale behind this is that it is necessary to keep a consistent supply of product manufacturers on line, through good times and bad. The theory is that if the government relied exclusively on the open market for its products and suddenly had a crisis where the government needed more product than the open market manufactured on a regular basis, the government would be stuck. A close friend wanted to share his insights and experiences with me concerning military spending, yet he did not want to divulge his name or rank.

> I am a program manager for the Navy. I procure all special Air Systems for the Navy and Marine Corps. My budget baseline for all my classified and unclassified systems exceeds $3 billion. I have also been a flight test guy (both operational and developmental), the Chief Pilot for the government at the Sikorsky, and have managed other large and small scale programs.
>
> For manned air vehicles the specifications are amazing and very difficult to meet, hence the $400 dollar toilet seat. Every piece of equipment in a navy aircraft has to be qualed to 20/20/10 … meaning 20 g loads in two axis, 10 g in the lateral axis before a flight clearance will be granted. Believe it or not, for transport aircraft the entire toilet assembly must be rated to those specs.
>
> Last, outside of the national lab structure, gone are the days when the government paced technology innovation … and the large defense contractors don't either. The hardest part of my job is maintaining software engineers writing code on my major programs: federal law only allows me to pay a certain wage for certain work … software engineers are so in demand that the gaming industry swoops up all my talent just when they get really good!

I spoke with another friend who has been involved in military and government contracts for over 30 years. He wanted to remain anonymous but provided the following scenario.

> Suppose you are serving in the military or better yet a contractor and are working on a large project and need a toilet seat. Not just any toilet seat, a special one that has to be a custom size and has to meet hundreds of different specifications with stacks of forms and documentation. You go to a company and ask them to supply that special toilet seat for you.
>
> The company assigns an engineer to design it and he determines it can be purchased at the local hardware store. He tells you this but you don't want to hear it, you want the company to supply it as you don't have an account at the store and don't feel like spending the next 6 hours filing out the documentation and forms yourself to get one. Besides, it will have to go through several offices and takes a few weeks to justify it. So, the engineer contacts the manufacturer and requests the specifications, weight, exact size, weight capacity, etc. All the information he

will need to submit to you for your forms. It all meets the specifications you laid out. He then fills out a requisition form and gives it to purchasing. They contact the hardware store and get a formal quotation, issue a PO, and the toilet seat is delivered. The bill is paid, end of story, right? No.

The engineer has spent at least 6 hours of billable time on your request. The purchasing agent has spent at least a half hour processing and getting the quote and the PO and processing the shipping costs. The warehouse guy spends 15 minutes receiving the seat, putting it on the receiving station, and calling the engineer to ask what he's supposed to do with it. Then he spends 20 minutes packaging it up (in the special packaging you required) and ships it to you.

The accounts payable clerk has another hour of billable time processing the check, entering the expenses in the log, and then billing you for all the costs incurred. In the end, that toilet seat has actually cost the company about $800 in billable time, but it's pencil-whipped down to $600 and the company ends up eating the extra cost, because the company doesn't want to upset you or lose you as a customer.

The press gets their hands on the story and incites an entire nation of uninformed, knee-jerkers to call for the company's head on a platter for supplying something to you at a loss. Now, if you had just gone down to the home depot yourself, and if you had gotten rid of most of your ridiculous red-tape and paperwork, this whole mess could have been avoided.

In the above scenario, where the bid is cost plus, the price per piece should go down dramatically as the number of units goes up. But, if the government buys one toilet seat for $4600 and then later that same year needs 100 more, they may go back to that company and order all 100 for $600 each, as that is the price that is set and accepted so it must be right. It happens. Now if that toilet seat went out for competitive bid and someone bid $50, it would probably be rejected because it is too low. They already know what one costs, $600. So the system will not allow a bid to come in so far under that price. There are types of bids (hard money bids) where the government decides beforehand approximately how much a unit should cost and will not accept a bid that is drastically higher or lower, no matter what. In some cases, they also will reject a bid that is more than a set percentage below the competing bids, and they think that if it is that low, it cannot be right.

Cost of Failure

When a failure on the production floor occurs, the effect on profitability is staggering, yet many organizations do not realize it. I witnessed this phenomenon many years ago at a bread manufacturing facility. During the production cycle, a sensor used to direct and divert loaves of bread malfunctioned. Dozens of loaves of bread began to quickly back up on the conveyor line. The line operator quickly stopped the line from moving and placed a call to maintenance. The maintenance technician was there in a matter of minutes, but by this time, hundreds of loaves of bread had been dumped onto the floor. This failure did not stop the loaves that were being removed from the large ovens and did not stop the hundreds of loaves from entering the oven. It also did not stop the hundreds of pounds of flour from being kneaded into dough and then divided into loaves.

While the technician fixed the sensor, the line worker was busy shoveling dozens of loaves of bread into large plastic carts. The loaves that hit the floor went into a scrap bin to be used to feed hogs, but the loaves arriving at the broken station were being put into a "seconds" bin that would be used to feed the homeless. There was nothing wrong with those loaves except that they were not being processed the same as the loaves destined for the grocery store. Soon the problem was solved. It took the technician several minutes to fix the sensor, and the line began to run smoothly again.

I asked the plant engineer what the downtime costs were when a situation like this occurs, and he said it was from $750 to $1500 per hour. That seemed very cheap considering all the people directly and indirectly involved in the failure. I then asked what he thought were the trouble stops on the line and what the overall cost of failure was. Here are his comments concerning the breakdowns:

> Conveyor line bearing problems from heavy loads, heat and friction as well as contamination in the form of water on the proofer and abrasive contaminants such as sugar and flour. When the conveyor fails it can take anywhere from 5 minutes to an hour to fix. Production stops and often parts are required. In the rare instance a replacement part is not readily available. When this happens a temporary line is set up. This can take 30 minutes to 2 hours to get going. The cost of this failure is over $50,000 annually which includes maintenance labor, the estimated downtime and the parts replacement.
>
> The proofer gearboxes all seem to have leaking seals from time to time, the motors burn out, we get gear and bearing wear, sometimes the water from the frequent wash down gets into the box and rusts the gears and mixes with the oil. Abrasive contaminants such as flour and sugar and different grains can do a job on the internal components. The wear and deposit build up can also take a toll on motors and bearings. We noticed that as a gear face becomes worn or deposits build up, the mating surface becomes unbalanced resulting in bearing stress and eventual failure. The motors will also experience higher operating temperatures resulting in premature failure of the windings. When this happens,

production stops. The total annual costs associated with these gearboxes are around $25,000.

Mixer hydraulics gives us problems as well. The flour, sugar and grains are a problem here and like the proofer gearboxes, the seals on the hydraulics leak. All the systems have severe seal leaks which seem to lead to contaminant entry and loss of fluid. This can also become a safety concern if someone was to slip and fall due to any oil on the floor but a greater concern would be the fact that the oil could (although unlikely) find its way into the mix. We use food grade hydraulic fluid but even then there can only be 10 parts per million allowed to contaminate. Any higher and the whole batch has to be scrapped.

I have noticed that the sumps are severely contaminated which can be directly related to leaking seals. The abrasive particles of sugar and flour enter the hydraulic fluid stream by a vacuum differential created upon thermal cycling of the fluid. It seems that when the fluid cools during off-shift, the volume retracts and brings abrasive particulate back into the sump where it will eventually migrate to the pump components leading to premature failure. When this happens, production stops. Fortunately this is not a common occurrence. All toll, these failures happen once every 4 to 6 months and account for roughly $30,000 in downtime, parts replacement and labor costs.

After listening to the various problems and the impact they had, it did not take long to realize that there was much more that these failures were costing the company than was first thought. Fortunately, the plant engineer and plant manager were open to various suggestions that would provide a quick return on investment (less than 3 months). The downtime, parts replacement, and labor costs associated with the common failures mentioned above have been eliminated. Other companies are not so open-minded and suffer the ravages of failure.

As stated earlier, there are fixed costs that remain constant, variable costs that increase with production, and profits that also increase with production. Graph 2.1 is a representation of a normal business operation.

Each asset within the company contributes to the expected profitability. When a new person is brought on board, there is justification for his or her salary. This is no different when a tool, a truck, or an injection press is bought. The capital investment was likely approved by upper management because there was a certain level of profit expectation. Unless a company is completely mismanaged, labor or capital assets will be accompanied by profit justification or a payback period. In some companies, the return on investment has to be as little as 6 months, while other organizations require up to 36 months for a breakeven point. When an asset such as an air compressor or a mill or even a hydraulic jack is no longer functional, the

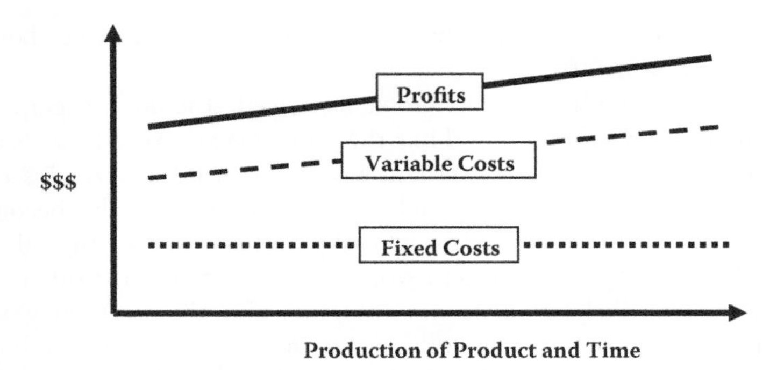

Graph 2.1 Normal manufacturing operation costs.

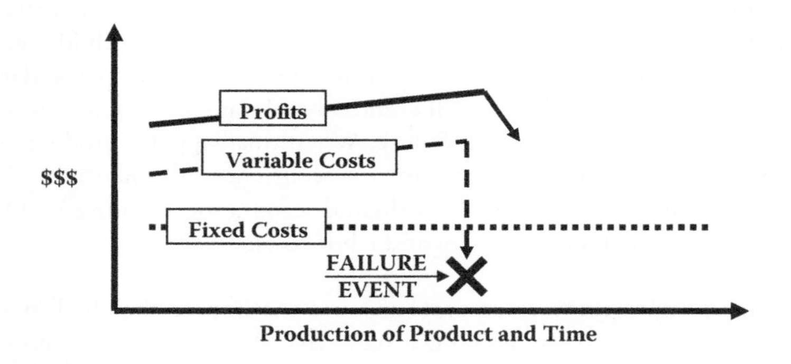

Graph 2.2 Profits, costs, and failure event.

expected production has been affected. That item has a contribution percentage to the overall profitability that is no longer being realized. Worse yet, any task that is directly related to the broken machine is now affected and considered inoperable until the failure has been relieved. Graph 2.2 represents a failure event. Notice how the fixed costs remain the same, there is a drastic reduction in the variable costs, and profits are on a downturn.

The sharp reduction in variable costs is typically due to the halt of direct material usage. Future profits are lost because no product can be made, although inventory can still be sold until it is gone. The fixed costs continue accumulating but are now wasted because no product is being produced. Variable costs will continue to fall because they are not used, whereas some, like maintenance, will suddenly rise in response to the incident. The cost for repair from a severe outage can be many times the profit made in the same time period. In the below scenario, the variable costs adjust to accommodate the failure event, the fixed costs remain, yet are

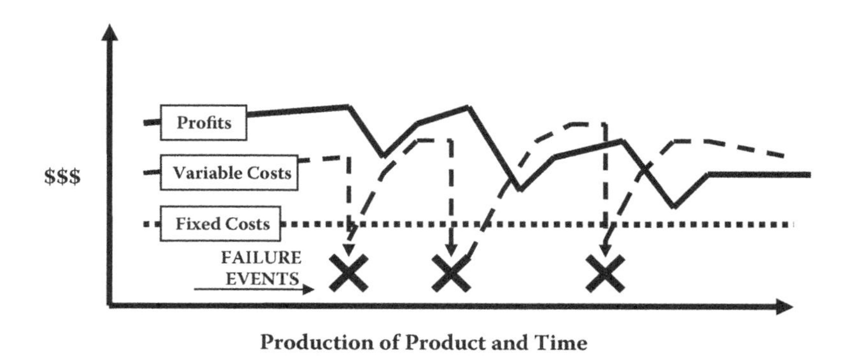

Graph 2.3 Repetitive failure events.

essentially wasted expenses due to a "no production period," and the profits adjust, yet have been diminished. Other failure events occur and continue to affect profitability. The variable costs adjust because more employees are getting involved and materials are being purchased. This activity continues to affect profitability. There are few companies that would consider the cost in time spent by maintenance, engineering, and purchasing to obtain replacement parts because of the failure. Realistically, these people would not be brought in if there had not been a failure. The cost is incurred only because the failure happened.

The failure event has now taken people away from their regular jobs and allocated them to the problem at hand. Another aspect of line shutdown is the various meetings associated with the event. Not only are these meetings a time drain, but they also contribute to bringing down the morale and spirit of the organization. At some point, if the failures are significant, there can be a profound effect on profitability and employee turnover. All too often a company will have to downsize, become acquired by another company, or shut down the plant (Graph 2.3).

Consider the following list of associated costs due to failure (and repeated failure) on a production line:

- *Labor (Direct and Indirect)*: Operators, repair technicians, supervisors, and management
- *Material Costs and Production Waste*: Scrap, replacement product, lost production, replacement parts, fabricated parts, shipping, storage, inventory replenishment, quality control
- *Services*: Emergency hire, subcontractors, travel expenses, consultants, utility repairs
- *Equipment*: Energy waste, start-up costs, shutdown costs, emergency hire, damaged items

- *Additional Capital*: Replacement equipment, new insurance spares, buildings and storage
- *Consequential*: Penalty payments, lost future sales, legal fees, loss of future contracts, environmental cleanup
- *Administration*: Documents, purchase orders, meetings, planning, schedule changes, investigation and audits, invoicing and matching, accounts payable/receivable

It becomes apparent very quickly that there are many people involved when failure strikes. It is also not surprising to understand that many companies and organizations do not consider the overall impact of failure. The act of purchasing an item that would have an influence 100-fold greater than the price is almost impossible to comprehend unless you were to consider how even the smallest, seemingly insignificant item can have such a profound effect. Consider that the next time you decide to ride a roller-coaster. Hopefully, the maintenance technician used the performance bolts built to a certain specification and not those from a local hardware store.

Success and reliability occur when failure can be minimized. Identifying potential failure sources and establishing proper protocols to offset these situations will result in dramatic cost savings. Successful companies will design products around failure or rather the elimination of potential failure sources as they relate to their particular product line.

Mantek LubraSystems is a relatively unknown company to the majority of factories and heavy equipment users but is quickly becoming a valuable asset to a growing number of mining concerns throughout North America. It is a company that develops and manufactures various lubrication products that are designed to increase equipment reliability while reducing the overall spending of parts and equipment components. One such application has saved INCO Stobie Mine in Sudbury, Canada, hundreds of thousands of dollars in reducing downtime, parts replacement, labor costs, and lubricant consumption. One application that Mantek LubraSystems was involved in was a reliability challenge to reduce maintenance costs of underground crushers. The rock crushers at the INCO Stobie underground mine are critical in the overall operation. Without the crushers, the rock would be completely unmanageable. When the rock is mined, large pieces are conveyed to be crushed. The rock or "muck" as it is called must be crushed to sizes no bigger than 7 inches so it can be brought to the surface for further processing. The rock crushers do the work of crushing the rock to a manageable size and experience incredible loads and contamination. The crushers are found from 1000 to 7000 feet below the surface. Component failure is routine and costly.

INCO has been in business for over 100 years and is one of Canada's best-known companies and largest exporters. INCO employs over 10,000 people around the world. The Stobie mine employs approximately 450 people. The mine produces copper, nickel, and precious metals such as gold, silver, and palladium. The Fuller-Traylor Double-Toggle Jaw Crusher is a critical piece of equipment that processes hard rock as large as 4 feet in diameter to quickly diminish it in size to 7-inch (or smaller) pieces.

Mines can be very unforgiving to heavy equipment. Heavy loads, friction, and contaminants can quickly degrade most lubricating greases, leaving the metal bearings unprotected and resulting in costly downtime, parts replacement, and additional labor. Before INCO switched over to Mantek's lubricating grease ELITE SUPREME, 48 manganese bronze bushings on the crusher had to be replaced annually at a cost of over $3000 each, totaling $144,000. The 4-ton toggle had to be replaced every 2 months at $10,000 each, which accounted for over $60,000 annually. Mantek LubraSystems was brought in to develop a reliability partnership that reduces downtime, lowers maintenance and operating expenses, and drives a proactive maintenance initiative. The objective included the use of a propriety bearing cleaning compound BEARING PURGE, a dry molybdenum compound DRI-GUARD PLUS™, and a severe duty lubricating grease ELITE SUPREME. Mantek National Account Manager Andre Leveke suggested cleaning the bearing while in operation by using a Mantek LubraSystems product called BEARING PURGE to flush contaminants from the Pitman and pillow block bearings while continuing to provide lubrication. A new grease was then used; ELITE SUPREME was introduced into the bearings as the BEARING PURGE was purged out. The cleaned bushings and toggle seats were treated with Mantek's DRI-GUARD PLUS, a dry film lubricant especially formulated with molybdenum disulfide to provide a long-lasting barrier under extreme pressure to help prevent metal-to-metal contact. The new grease, ELITE SUPREME, is capable of withstanding extreme pressure and high heat as well as being very resistant to water washout even when a power washer was introduced.

Flushing out the Pitman bearing and pillow block bearings with BEARING PURGE and switching to a higher-performing grease reduced operating temperature over 19°F within the first 3 hours of operation. Free of contaminants, the glide time on the Pitman has quadrupled with the new grease ELITE SUPREME. Since October 20, 2005, there have not been any lubricated-related downtime incidents. There have been no bushings or toggles replaced. Maintenance Foreman Marcel Garneau of the Stobie Mine happily reported that he now has 64 more man hours per month freed up due to increase reliability. Monthly grease consumption has dropped from 620 kg of ESSO Unirex™ EP2 monthly to 130 kg of ELITE SUPREME. Since using the Mantek LubraSystems products, bearing repairs have been reduced from monthly replacements of the bearings costing over $12,000 to bearings lasting over 6 months, and toggle replacements were being replaced twice annually at a cost of $20,000. The labor costs were $2240 or $27,000 annually. These costs have been dramatically reduced, and operation savings of $167,000 annually is a conservative estimate. The savings on reduced lubricant usage combined with a reduction in labor, parts replacement, and downtime could easily equate to over half a million dollars a year. Translating that savings to other operations could help INCO save millions of dollars on operational costs.

Case Studies—An Effective Way toward Change

Case studies are one of the most effective ways to understand the value associated with changing to a new tool, device, or material. The work that INCO and Mantek LubraSystems did sensationally was a case study. Mantek's goal was ultimately to capture the entire business that INCO had to offer. INCO was not about to roll over and just hand it to them. The business had to be earned and value sustained. Many companies in the past would call on INCO with various solutions to their problems. The difference with Mantek LubraSystems was that the solutions suggested were developed with reliability and cost reduction in mind. Once the technology is proven sound in several severe applications, it would stand to reason that further value can be obtained in other applications. Often, such a change can take on a life of its own. I spent many months going through General Electrics Six Sigma training. Six Sigma is essentially a process improvement process with a goal of reducing the opportunities for failure to below three for every million chances—basically running an error-free process. The tools used in establishing a Six Sigma process can be applied to case studies provided adequate resources are available. The process is very logical and can be applied to most any engineering change, product, or process evaluation or even product or process development.

There is a misunderstanding that a Six Sigma process requires complete "buy-in" from upper-level management and a Six Sigma master black belt to champion the cause. Nothing could be further from the truth. In actuality, the Six Sigma process is very logical, and if the parameters and goals are kept in check, they are actually rather simple to follow.

Step 1 Define: Define exactly what the problem is, why it is important, who is involved, and what the expected outcome target is.

Step 2 Measure: Decide on how the change is going to be measured. Make sure that there are several criteria being evaluated and the measurements actually have relevancy to the operation costs. This is sometimes lost on some projects.

Step 3 Analyze: What are the sources of failure in terms of the effect on downtime, parts replacement, labor rates, and productivity? Looking at all the failures in a certain time frame and assigning their cost influence, you will be able to decide on which has the greatest influence and needs to be addressed.

Step 4 Improve: Implement a solution in terms of a new product, process, or material.

Step 5 Control: Once the change is in place and sufficient measurements are recovered, it is now time to examine the benefits of the change in terms of improved processing and cost avoidance.

If one followed the above process yet did not identify it as a Six Sigma process, it would be considered the foundation of an excellent effort in operational improvements. The Six Sigma process improvement methodology may be effective but can convolute the situation if taken to an extreme. It is unfortunate that many companies have tried to apply the strict code of Six Sigma statistical analysis only to bring the effort to a screeching halt. A simple case study using these principles is often enough to explore new ideas and possibly reduce operational costs. A case study provides the user and the supplier with such confidence. The elements of the case study are based on observation and documentation with an effort toward keeping as many variables that would affect the results to a minimum. The structure of a case study using the aforementioned Six Sigma tools is as follows:

Problem: What are the specific problems as they relate to equipment and process?

Proposed Solution: How will the new tool, material, or device solve these problems?

Study Layout: Measurements are taken before, during, and after the trial.

1. What are the conditions being monitored (i.e., temperature, pressure, energy in amperes, RPMs, throughput, etc.)?
2. How are the measurements going to be taken?
3. What is the frequency of the measurements?
4. What is the duration of each segment?

Results: Document the results during the process and before, during, and even after the changeover.

Data Interpretation: How did the implementation of the new tool, material, or device affect the equipment, and what does that mean in terms of energy costs, downtime reduction, reduction of parts replacement, and labor cost reduction?

Keep in mind that data derived from a controlled test may not be indicative of an actual trial, and an actual trial may not be conclusive of the overall performance, yet to justify the pursuit of a new methodology, overall performance is driven by a trial, and any trial must be derived from controlled test data.

Other Costs Defined

Administrative Costs: All executive, organizational, and clerical costs associated with the general management of an organization rather than with manufacturing, marketing, or selling.

Common Costs: A cost that is common to a number of costing objects but cannot be traced to them individually. For example, the wage cost of the pilot of a 747 airliner is a common cost of all of the passengers on the aircraft. Without the pilot, there would be no flight and no passengers. But no part of the pilot's wage is caused by any one passenger taking the flight.

Conversion Cost: Direct labor cost plus manufacturing overhead cost.

Cost Behavior: The way in which a cost reacts or responds to changes in the level of business activity.

Cost Object: Anything for which cost data are desired. Examples of possible cost objects are products, product lines, customers, jobs, and organizational subunits such as departments or divisions of a company.

Cost of Goods Manufactured: The manufacturing costs associated with the goods that were finished during the period.

Differential Cost: A difference in cost between any two alternatives. Also see *Incremental Cost.*

Differential Revenue: The difference in revenue between any two alternatives.

Direct Cost: A cost that can be easily and conveniently traced to the particular cost object under consideration.

Direct Labor: Those factory labor costs that can be easily traced to individual units of product. Also called *touch labor.*

Direct Materials: Those materials that become an integral part of a finished product and can be conveniently traced into it.

Fixed Cost: A cost that remains constant, in total, regardless of changes in the level of activity within the *relevant range*. If a fixed cost is expressed on a per unit basis, it varies inversely with the level of activity.

Incremental Cost: An increase in cost between two alternatives. Also see *Differential Cost.*

Indirect Cost: A cost that cannot be easily and conveniently traced to the particular *cost object* under consideration.

Indirect Labor: The labor costs of janitors, supervisors, materials handlers, and other factory workers that cannot be conveniently traced directly to particular products.

Indirect Materials: Small items of material such as glue and nails. These items may become an integral part of a finished product but are traceable to the product only at great cost or inconvenience.

Inventoriable Costs: Synonym for *product costs.*

Manufacturing Overhead: All costs associated with manufacturing except *direct materials* and *direct labor.*

Marketing or Selling Costs: All costs necessary to secure customer orders and get the finished product or service into the hands of the customer.

Opportunity Cost: The potential benefit given up when one alternative is selected over another.

Period Costs: Those costs that are taken directly to the income statement as expenses in the period in which they are incurred or accrued; such costs consist of selling (marketing) and administrative expenses.

Prime Cost: Direct materials cost plus direct labor cost.

Product Costs: All costs involved in the purchase or manufacture of goods. In the case of manufactured goods, these costs consist of direct materials, direct labor, and manufacturing overhead. Also see *Inventoriable Costs.*

Raw Materials: Any materials that go into the final product.

Relevant Range: The range of *activity* within which assumptions about variable and fixed cost behavior are valid.

Schedule of Cost of Goods Manufactured: A schedule showing the direct materials, direct labor, and manufacturing overhead costs incurred for a period and assigned to *Work in Process* and completed goods.

Sunk Cost: Any cost that has already been incurred and that cannot be changed by any decision made now or in the future.

Variable Cost: A cost that varies, in total, in direct proportion to changes in the level of activity. A variable cost is constant per unit.

Determining the Return on Investment (ROI) of a Purchase

To determine the ROI of an item, several criteria must be understood. ROI is considered a key financial metric for industrial capital expense as well as personal investments. It is the ratio of the net benefits over the cost. There are two fundamental methodologies that have wide acceptance in business and personal finance through which one can conduct an ROI analysis. They are the discounted cash flow analysis (DCFA) and payback period analysis (PPA).

Payback can occur directly and indirectly. The direct, profit-generating contributions of a part, tool, new technology, or process, as well as the indirect measures valued by management should be considered when calculating the ROI. The discount rate also known as the weighted average cost of capital (WACC) is the opportunity cost of the expected rate of return that could be obtained from other similar items or processes. This sort of analysis is beneficial when determining if a new part, tool, or MRO item should be considered. Often a case study provides essential information that can be translated plant-wide. If a particular item demonstrates a degree of savings in terms of reduced downtime, parts replacement, labor costs, and consumption, then those results can be magnified and significant costs savings (hence, profit generation) realized.

Another term often used when determining ROI is the net present value (NPV). It is defined as the difference between the cost of an item or investment and the return measured in same-day dollars. NPV calculations account for money's time value by discounting the future cash flow of the investment at some rate that varies with the risk of the investment. The NPV calculation determines the present value of the return and compares it to the initial investment:

$$\text{NPV} = \text{Initial Investment} + \text{Present Value} \tag{2.4}$$

If a hydraulic pump costs $20,000 and will save in downtime, part replacement, and labor $5000 per year for 5 years, there is a $5000 net return on the investment. The NPV of the investment, however, is actually less than $5000 due to the time value of money:

Net Present Value = [Net Cash Flow for Year 1/(1 + discount rate)] + [Net Cash Flow for Year 2/(1 + discount rate)] * 2 + … + [Net Cash Flow for Year N/(1 + discount rate)] * N

$$NPV = \left[\frac{NCF1}{\frac{1}{DR}}\right] + \left[\frac{NCF2}{\frac{1}{DR}}\right] * 2 + \left[\frac{NCFn}{\frac{1}{DR}}\right] * n \tag{2.5}$$

However, if there is a purchase that requires and produces a number of cash flows over time, the internal rate of return is defined to be the discount rate that makes the net present value of those cash flows equal to zero. In other words, the discount rate that makes the project have a zero NPV is the internal rate of return (IRR). The IRR method of analyzing a new item or using a technology in a project allows a company to consider the time value of money. IRR enables you to find the interest rate that is equivalent to the dollar returns expected from the technology or project under consideration. Once a company knows the rate, it can compare it to the rates that it could earn by investing money in other technologies or projects or investments. The ROI is just a percentage, so include the payback time to make it persuasive. If the hydraulic pump purchased was $20,000 and generating $40,000 a year in profit, it paid for itself within 6 months. Costs divided by monthly benefits yields the number of months to payback.

The processes by which one can determine the ROI of any purchase or investment are very simple. This information will become essential as a file for a new item is established. Many purchasing managers as well as production and maintenance managers will use this information as backup information for decisions made.

***Step 1*: Calculate All the Costs Associated with the Purchase**

What are the initial upfront costs?
What are the maintenance costs?
Are there any fees or taxes?
What kinds of costs are there to properly evaluate the purchase?
How much of your time will this evaluation take?
Once you do one, all others become very simple. Often the same values can be copied over and over.

***Step 2*: Estimate the Results of the Purchase**

How much do you expect to gain?
When do you expect this to happen?

***Step 3*: Establish a Timeline for Costs and Returns**

Draw a simple timeline or just list in chronological order all the costs and returns you discovered in steps 1 and 2. Costs should be listed as negative dollars and returns as positive dollars.

For example,

April 15, 2007: Hydraulic pump cost: $100,000
September 1, 2007: Production improvement: $120,000

***Step 4*: Calculate the ROI**

In order to understand the ROI, you must measure the amount of reduction in operational costs. The ROI comes from the increased operational efficiency and reduced costs achieved by reducing the repair and replacement of components as well as a reduction in equipment downtime and labor.

Factors to be included in ROI calculation include:

Costs and Expenses
- Additional hardware cost (*a*)
- Software cost (*b*)
- Training cost (*c*)
- Retooling cost (*d*)
- Monitoring cost (*e*)
- Operational costs from vendor consulting (*f*)
- Additional costs (*n*)

Technical Benefits
- Reduced downtime cost (*A*)
- Reduced labor cost (*B*)
- Reduced component replacement cost (*C*)
- Additional Savings (*N*)

There may be other direct and indirect expenses and cost benefits including increased process efficiency and reduced scrap. These also have to be accounted for in your ROI calculation.

Applying the ROI Formula

Once all the costs and expenses along with the benefits have been understood, it is time to apply the numbers to the ROI formula:

1. Calculate the total cost of implementation. Sum up all the expenses that we listed from point *a* through *n*.
2. Determine the total savings resulting through the technical benefits by going through points *A* through *N*.
3. The ROI is essentially the sum of the benefits *A* through *N* minus the sum of the costs associated *a* through *n*:

$$ROI = \sum A + B + \ldots N - \sum a + b \ldots + n \quad (2.6)$$

A spreadsheet is also a simple way to carry out the analysis. Many include a great function to find the IRR. The Microsoft® Excel® (MSFT), Open Office® (SUN), and Google® (GOOG) spreadsheets include the XIRR function that can calculate the ROI.

By having the costs associated with the purchase, the result of the purchase, a timeline of the costs and benefits, and the expected ROI, the purchasing process now becomes very easy. The decisions made will have a lasting effect on the bottom line and the overall operations. To carry out this function, costs and value must be understood before a specification can be put in place.

Summary

At this point, you should have a firm grasp of performance and quality. There is a difference, and in most all circumstances, choosing a product that has higher levels of performance and performs consistently at those levels will help drive down costs. At this point, you should also be aware of warranty concerns, product recalls, and false claims and be aware of your rights. Operation and time costs are important elements that should now be clear to you. From that, you should also understand the concepts of a profit return from a purchase. The items that you will eventually specify will have an impact on the bottom line. You will not just be "placing orders" but rather driving productivity. This can only come about through an understanding of the cost of failure. Failure can be a valuable asset if used to eliminate itself. From that, calculating your return on the purchased item will follow.

In Chapter 3, you will be introduced to various sources of failure. This information is extremely important and valuable when interviewing technicians as well as poring over technical data sheets. Having a good grasp of failure sources will help establish a strong foundation for a performance-based purchasing specification.

Tools for the Specification Development Process

Return on Investment (ROI) Work Sheet

Costs and Expenses	*Item 1*	*Item 2*
Product cost (*a*)		
Ancillary product cost (*b*)		
Training cost (*c*)		
Retooling cost (*d*)		
Monitoring cost (*e*)		
Operational costs from vendor consulting (*f*)		
Additional costs (*n*)		
Total Costs and Expenses		
Benefits and savings		
Reduced downtime cost (*A*)		
Reduced labor cost (*B*)		
Reduced component replacement cost (*C*)		
Additional savings (*N*)		
Total Amount of Benefits and Savings		
Determining ROI $(A + B + C + N) - (a + b + c + d + e + f + n) =$ ROI		
Benefits and Savings minus Costs and Expenses		
Breakeven Point		

Notes: The values for the *Benefits and Savings* may have to be determined on a monthly scale. Once done, the values can be multiplied by the number of months (or days) it will take to get to a balance against the *Costs and Expenses*. After that balance is met, the ROI will begin to emerge.

Chapter 3

Source of Failure

In this chapter, the root causes of failure will be determined. Understanding why a tool, component, or part fails is essential for the development of a comprehensive purchasing specification. Once failure is understood, performance criteria can be established. The goal of Chapter 3 is to:

1. Understand and define *failure*
2. Identify failure modes from man, method, material, and machine
3. Understand the fundamentals of a documentation method
4. Begin to fill out the total cost work sheet to determine which items to focus on
5. Begin to fill out the failure mode work sheet to understand the reasons for failure; this is the beginning of the specification development process

Assessing the Situation

> If I walked in a one factory I walked in a tousand. I could tell yas witin fife minute if da place was go'n to make it or close its doors in a few years just by the feel of it ... the sound it made and how the line moved. Like go'n to a swing hall. Some folks like swing halls for da dancing and tunes dat is. Dey feel dat beat pound'n away ... reminds dem dey're alive! Dese places reminds me of dat. Remind me dat da whole town is still alive heck maybe even the whole country. But if da place sounds and feels sad den it don't got long and dat's about as sad as it gets. Ya know a good blues man's can make yo cry cuz he's gots is pain and ee's turning it. A factory does da same I tell ya. Dose lines and

> presses scream o hum and swing o sway. Dey always make'n a song too doncha know unless dey dead dat is. But dat song gots to be sweet if its gots a chance. You go now, you stand still ova dair in dat quite corner of da shop dair, off aways and yo tells me yo if'n yo feel da buzz of da floor right true yo boots. Dis place gots one. Every place is gots one ... a rhythm dat iz and sometimes it makes you smile, mmm ... mmm and other times it jus makes yo feel like cry'n ... mmm ... mmm ... mmm.
>
> From the author's unpublished, private novel *American Standard*

On any given day, I am granted the opportunity to meet with maintenance managers and supervisors from all sorts of businesses. Many of them have worked their way up through the ranks to take on leadership roles, and some were transferred or promoted in. I found an interesting difference in the group leaders with a certain amount of "technical chops" and those without. As prior mechanics and technicians, they had the opportunity to be critical and provide suggestions they felt would make a difference in the way the department was managed or the way the company was run. The combination of mechanical ability, leadership quality, and ambition were characteristics influencing their promotions. Once leadership roles were assumed, many newly appointed managers quickly realized that there was much more work and strategic decision making involved in the job than previously understood. The proverbial shoe was now on the other foot. The ones that succeeded did so because they blended a combination of decisiveness, experience, empathy, and organization into the job. The managers and supervisors who spent a certain amount of time down on the floor understanding problems that occurred and listening to the operators and mechanics (and even the machines) were the ones who became the truly outstanding managers. The idea of spending time understanding the problems by direct involvement is critical and beneficial regardless of the machine or application. A maintenance manager needs to spend time on the floor with those who are actually skinning their knuckles and getting dirty. If not, they will never truly know where the problems are and how to effectively address them.

When our son Max was 5 years old, he was having problems getting his shoes on. Of course, this was right before he was to get into the car to go to school, so the anxiety of being late only compounded the problem. It was not until I actually got down on the floor and tried to understand what the problem was that I was able to provide a solution. First, he had tucked the tongue of his sneaker under his foot as opposed to letting it rest on top. Second, he was attempting to loop the ends through and not just tying the loops. It was only after I spent time "on the floor" was I able to understand the problem. I offered up some quick and accurate suggestions that would ease the anxiety and "increase his production rate." Problem solved; no tardiness. Max is happy, Dad is happy, and most importantly, Mom is happy.

Problems that occur on the production floor are far more complicated and critical than understanding why a 5-year-old is having problems tying his shoelaces. But the idea of actually spending time understanding the problems by direct involvement is critical and beneficial regardless of the machine or application. You have to spend time on the floor with those who are actually skinning their knuckles and getting dirty. If not, you will never know what the problems are and how to effectively address them.

The role of the maintenance technician has evolved over the past century. Many years ago, teams would be relied on only to fix a machine when it broke. This approach is known as "reactive maintenance." Even today, this approach is common mainly because many companies just cannot catch up to do the types of things they know would actually increase reliability and decrease downtime. Instead, they only attack the immediate need of the production floor. In many cases, there is not enough time or resources to engage in preventative, proactive practices.

Shortly after World War II, there was an increase in economic prosperity. An increase in domestic production of various items followed. The increase in production demands meant that equipment had to produce more products at less cost. For the first time, the use of electronics was heavily relied upon. This had a direct effect on productivity, but it also meant that maintenance technicians had to become quick studies in troubleshooting these devices. Computers (albeit first-generation punch card and reel) were now being used by many of the forward-thinking facilities for record keeping and data processing. Data were stored and analyzed. Manufacturing and maintenance technology in the late 1950s blossomed due to competition and innovation. Even the "space race" had an influence on manufacturing and maintenance. Large data management allowed for comprehensive preventive equipment maintenance techniques to be born. A new age of maintenance and reliability was established with a direct impact on productivity.

During the space race, the National Aeronautics and Space Administration (NASA) found it necessary to employ a technique called failure mode effect analysis (FMEA). Aerospace technology demanded that equipment failure was not an option. Therefore, when a part broke, the root cause had to be identified and a corrective action (and potential design changes) had to be put in place. This approach was quickly adopted by many factories, including part, tool, and machine manufacturers. FMEA had a direct effect on the bottom line and the quality initiative that many companies were attempting to develop. This technique helped foster in users of the next generation of maintenance approaches—*preventative maintenance.*

Among many things, preventative maintenance is a practice of scheduling the rebuild or replacement of parts or machines prior to failure. Through record keeping, it is possible to determine that certain parts will fail after many hours of operation. Scheduled downtime to administer preventative maintenance became routine, with various parts being rebuilt or replaced prior to failure. It became quickly understood that unscheduled downtime could be reduced dramatically.

Using preventative maintenance techniques, the maintenance team could now contribute to the facility's efficiency and profitability. Maintenance departments transitioned from a lost cost center to a loss prevention center.

As preventative maintenance techniques and practices began to flourish and evolve, the next natural progression would be to have a technician attempt to predict equipment failures as well as try to prevent them. This form of mechanical clairvoyance is actually well understood. Many years ago, it was not uncommon for mechanics to listen to a bearing or a gearbox with a screwdriver. The mechanic would put the screwdriver handle to his ear and place the tip onto a gearbox, pump, or bearing housing and "listen." This is a technique I use to this day on various pieces of equipment, including engines. The human ear is especially sensitive to various vibration states, and our brains are wired in such a way that we tend to remember distinct variations of different sounds. It is actually rather easy to pick up the slightest amount of grit grinding away at a bearing race or a pump experiencing subtle cavitation. Many companies have taken this concept and explored the nonaudible regions of sound (ultrasonic analysis). This technique does not stop at sound but reaches into the broad spectrum of vibration (vibration analysis) and on through to the electromagnetic spectrum of heat (infrared analysis). This approach is known as "predictive maintenance and condition monitoring," when various analytical techniques are done on a regular basis to determine if failure is approaching. In this manner, action can be taken if failure is imminent. If preventive maintenance is a good diet and exercise, then predictive maintenance and condition monitoring are a checkup by your doctor that includes blood work (oil analysis), listening to your lungs and heart (acoustic and ultrasonic equipment analysis), taking your temperature (thermal analysis and infrared imaging), and a stress test (vibration analysis). These are only a few concepts being implemented today with great success.

Failure Defined

To understand the performance requirements, one must first identify the cause of downtime, parts replacement, and labor costs. In order to establish performance criteria, it is essential to determine how something fails to perform. How does a product fail?

Failures are:

- Any loss that interrupts production.
- A loss of asset availability.
- The unavailability of equipment.
- A deviation from the status quo.
- Not meeting target expectations.
- Any secondary defect.

Several techniques can be used to identify a failure mode and root cause of a problem. FMEA is a process used to identify potential failure modes and determine the effect of each on the overall system performance and cost. The process is broken down into four areas:

- *Potential Failure Modes* (One or a Combination of the Following)
  - Materials
  - Manpower
  - Methods
  - Machines
- *Potential Failure Effects*
  - What do we do when failures occur?
  - How does it affect the customer and us?
- *Potential Causes*
  - What happened to cause the failure?
- *Current Controls*
  - What can help to prevent the failures?
- What is the process to prevent failures?

Examining the answers to these difficult questions can bring about improvements. Utilizing them in the development of procurement specifications can increase plant reliability with dramatic results.

Failures can be hard to categorize. They can be a by-product of negligence, or they could categorize with natural processes. Failure can happen due to negligence. If an overseer failed to inspect or repair failing equipment, then the overseer is guilty of negligence. Failing to notice and fix such problems could lead to a broken part or component or serious accident. A failure can also occur due to natural processes. Part or component rust, corrosion, and oxidation are just a few examples of failure due to natural causes primarily due to the environment. Stress fractures, elongation, cracking, shrinking, or warping are a few examples due to the working environment. Once the failure is identified, it is vital that it is documented. There are two main sources of failure information—from the equipment maintenance database or from interviewing the maintenance and engineering staff. The technicians will give certain information that will not be captured in the database. In many instances, the equipment they work on has particular quirks and operating characteristics that are never captured in the maintenance logs. Equipment is often compared to children. Each one is unique and special. Even twins have different personalities and needs. Consider the example of two presses. They are the same make, model, and year and have the same hours, and yet they will perform differently without explanation. Only the operators and maintenance staff are aware of any issues, and many times this information is not captured. Once the failure mode is understood, comparisons of improved performance can be analyzed.

Taking Advantage of Failure

A small appliance manufacturing facility was purchasing an adhesive to bond a handle to a toaster-oven door. The adhesive they had chosen passed all the engineering design criteria and was offered at a competitive price from a local distributor. The distributor only stocked the adhesive in 8-ounce tubes. The tubes were offered 10 tubes to a case (5 pounds worth of adhesive). The plant went through approximately 500 pounds of adhesive per month (100 cases of adhesive). The time it took to load the adhesive dispensing guns and the additional work required to dispose of the empty tubes was never considered a factor when the design team selected the adhesive and the purchasing department bought the adhesive. It was only when the product-engineering department was given the task to examine all parts of the product process that the indirect cost of the adhesive become apparent. This was after the product had been in production for 7 months. The distributor tried to negotiate a lower price in order to save the business. The difference of the prices still did not offset the cost of the additional labor. A competitive vendor approached the company with the concept of bulk dispensing from 450-pound drums. The distributor even offered to provide the bulk dispensing system at no cost provided the customer signed a multiyear purchasing agreement. The price of the bulk adhesive was 28% cheaper, and the company had increased labor savings due to the improved process. The bulk system also provided better control of the dispensing and less waste. The tubes always contained at least an ounce of adhesive after they were thrown away. They also provided an extra drum on site at no cost until opened. This was done just in case production quotas increased and the adhesive had a lead-time from the plants. The competitive distributor obtained the business because the distributor was able to offer the product in a package that helped improve savings.

Many vendors offer special services to their key accounts that they normally would not share with a small to medium-size customer. This does not have to be the case. An example of such a service would be the story of the airplane manufacturer and the safety shoe vendors.

Many years ago, a large company that built airplanes was experiencing a considerable amount of worker disability claims. The majority of these claims came from workers who were falling from the wing structure. Upper-level management became involved when the cost to deal with the claims exceeded a particular level. Purchasing rallied the three safety shoe vendors and asked for help. All three companies (A, B, and C) were given the task of coming up with a shoe that provided greater traction. Upper management believed that greater traction would mean less slip—less slip meant lower worker disability claims. The airplane manufacturer would even contribute to the development of a better sole. Companies A and B had approximately 45% of the business each. Company C had barely 5% of the business, mostly because their shoes were 20% more expensive than A or B. A and B came back with a sole design they felt would provide improved traction. No guarantees were made. It would take 6 months to determine the impact and benefit of the new soles. The development costs were considerable, and there were no guarantees it would work. Company C took on the challenge with a different approach. Company C understood that there was a safety concern but did not want to only consider the soles. For a minimal cost, they sent in a few engineers who specialized in materials, risk management, and occupational safety. They determined that the real issues were not with the soles of the shoes but with inadequate safety harnesses and a lack of safety training and protocols. They promptly presented a corrective action plan. Once implemented, they were able to reduce accidents by 95% almost immediately. The savings from the reduced worker disability claims offset the entire cost of outfitting all the employees with new safety shoes from company C. The airplane manufacturer was impressed. They gave company C 3 months to prove it. Company C actually saved the plant 15% more than estimated. They were awarded the entire safety shoe contract worth several million dollars a year. Company C had the resources to provide the solutions that took into consideration the direct and indirect costs of safety breakdown. They provided value-added contributions that helped drive productivity and cut costs.

Sources of Failure: Man, Materials, Methods, Machines

Anyone who has ever played organized sports or served in the Armed Forces had the following statement drilled into their heads: "Failure is not an option." This statement is true—failure does not have an option … of occurring. Failure will occur at some point; it is merely a question of when. There are some people who believe everything happens for a reason, and then there are those who believe there is a reason why everything happens. Whether you believe in divine design or random occurrence, things fail. Success and reliability occur when failure can be minimized. In general terms, the causes of failures can be grouped into four categories—man, machine, method, and materials:

- *Man* as a failure source can occur due to lack of training or experience, human error, or poor supervision.
- *Machine* as a failure source can occur due to poor maintenance, equipment used beyond its design criteria, or machinery that is old or outdated.
- *Methods* as a failure source can occur due to lack of supervision, no standard appropriate operating procedures in place, or lack of management skills.
- *Materials* as a failure source can occur due to using wrong materials of construction, external influences of contamination, or improperly manufactured materials.

Once these categories are identified, then a root cause can be identified and corrective measures put into place. If done so successfully, the opportunity for that failure to occur again will be reduced dramatically. If maintenance management can get a clear understanding as to why various pieces of equipment fail, then implementing corrective actions for improved reliability will follow.

Manmade Failures—Failure of Employees

In any given factory or shop on any given day, there is a fantastic dynamic at work that spells success or failure. If the business model is sound and management is not contributing to the failure of the business, then attention should be paid to the employees. We have all worked with someone who is a drain on the company. Indeed, we often wondered how they ever were hired and, worse, why they were still around. The unfortunate fact is that the odds that these folks will ever change are remote. There may not be much one can do about these sorts of failures, but they should be brought to light nonetheless.

We all know these types:

The Royal Highness: This employee blue blood believes that some, if not all, of the tasks required to do the job are beneath him or her. In certain instances, the employee will try to pawn them off on an unsuspecting Doofus and then provide total blame when things go bad. Often the employee has a very high opinion of himself or herself, yet it is all smoke with nothing burning. Most have little talent; in rare cases, the employee combines some Scammer techniques and works his or her way into a management role. Most everyone who has ever worked with the Royal Highness realizes what he or she is, and if a management role is ever assumed, it is short lived.

The Scammer: This rascal is shrewd and cunning and can often outperform everyone else in the department if focused on the job at hand. Unfortunately, this employee spends far too much time and energy scheming and defrauding the business or planning ways to manipulate other employees into doing something for his or her own self-interests. The Scammer gains satisfaction in "getting over" on the company. Scammers are very competitive and view management and upper management as adversaries; anytime this employee gets something by management, the employee feels he or she has won. When directed properly and compensated appropriately, this employee can be a star. Often this employee does what he or she does for the challenge and excitement and keeps doing it for the self-perceived rewards.

The Gadfly: This insect is so deeply absorbed in the office politics, business rumors (about the competition or his or her own company), or the various assignments that other employees are working on that the employee loses focus on what his or her job is supposed to be. The Gadfly has talent and passion, but they are misdirected. Often this employee is not challenged enough. With aggressive management styles and increased workload, the Gadfly can succeed. Caution–the Gadfly might be a Chronic Ache in disguise. The Gadfly is difficult to discern, and sometimes a metamorphosis occurs. Once this change occurs, it becomes a manager's nightmare—The Gadfly combines rumor, myth, and talent to become an amazing burden.

The Doofus: This employee simply cannot think. The Doofus often hurts himself or herself or breaks equipment, cannot remember various tasks, and requires almost constant supervision. It is unclear how this employee got to the position he or she is in, because if the employee were interviewing for his or her job, he or she would not stand a chance. Often this employee is friendly and would do anything you ask. The employee will work long hours and rarely complain. The problem is most of the employee's work could have been done in a fraction of the time. He or she is a drain on the company and a burden on the team and on his or her boss. Many times, it is believed the Doofus either is a substance abuser or has legitimate mental or psychological problems—nobody could be like that without a good explanation.

The Chronic Ache: This pain produces more lost time and contributes to the overall downfall of the group or department more than any other employee. The Chronic Ache will complain about everything and make it very difficult for other employees to remain positive. This employee believes the company owes him or her and has taken advantage of him or her. There is a recognizable gray mist of discontent that follows this employee wherever he or she goes, and this employee can suck the life out of anyone with the bad fortune of asking about his or her day. This employee is typically shuffled around to various departments. A boss may take the employee on as a test of his or her management mettle, or a boss may inherit the employee because the boss was blindsided or railroaded into taking the Chronic Ache.

It can be difficult on management to deal with these various types. Often, employees exhibit only nuances of these characteristics, and in some cases, they can work their way through them.

"I see employee failures every day, little failures hopefully," said Chris Aldrich, vice president of Mechanical Drive Components Inc. "It is a necessary part of the learning process. I think employees fail on a grander scale because of inexperience. Life is a series of failures and reactions to failures. It's what you gain from failure that makes you a real person. Often a company has to put forth the right training, as well, rather than give lip service" (personal communication, April 13, 2008).

Sometimes the employee is actually overemployed. Their skills or experience do not match the job description—a bad fit. They could also be underemployed, bored, burned out, or depressed. They may also have the martyr syndrome, thinking they work harder and longer than everyone else does, but still cannot get their job done.

Alternatively, they have mastered the passive-aggressive behavior. "Oh, nobody told me about that." There are even those who are perfectionists, who can never complete anything because it is never quite good enough. They are afraid of failure. Then there are the procrastinators. These employees wait until the last minute to start projects. Fortunately, procrastinators usually seem to always pull it out in the end. (I should know.)

Manmade Failures—Failure of Managers

If physics is the science of pushing matter around, then management is the science of pushing people around. With physics, great things can be accomplished. With management, business can flourish. Anytime a group of people is assembled to complete a task, and coexist, management is required. Families, sports teams, military forces, and production staff all require management. But managers have to be well organized, well trained, have the subordinates' best interests at heart, and maintain a steady driving force. In reality, that is not how it always works.

Often, management is the sole reason a company fails. Not unlike an unbalanced bearing or a hot running motor, management failure can be predictable, and the root cause can be understood. Of course, understanding why a motor fails is far easier than understanding why a manager has failed.

There are loads of reasons why management failure occurs, but bad managers can basically be characterized into four groups. (The really bad ones may actually belong to two or more.) They are as follows:

The Roach: The manager who lacks integrity. This creature is greedy; may look to capitalize at the expense of its subordinates; can be dishonest, immoral, and unethical; may lack accountability; may be very inwardly fearful; and may have a huge displaced ego. This manager believes that his or her market understanding and vision supersede the corporate objective. The manager also knows how to play upper-level management. The Roach typically throws subordinates under the bus while offering up solutions that may actually have been suggested by another subordinate. The chance of the Roach giving credit where it is due is very unlikely. Of the four categories, the Roach has the greatest survival skills.

The Hack: The manager who lacks ability and knowledge as it pertains to the business or subordinate management skills. The Hack possesses little, if any, market intelligence. Traits also include a lack of organizational skills, follow-through, follow-up, and an inability to motivate or inspire subordinates. The Hack may have difficulty retaining or even recruiting the right subordinates. This manager often micromanages or overanalyzes to the point of paralysis. The Hack may ask for input and ideas but never implement any of them. An interesting defense mechanism is a creation of bureaucracy so that ideas and decisions have to go through numerous approval processes. The inability to foster a team or provide adequate training or mentoring while not setting clear goals and expectations define the Hack. The Hack can also be critical of categorizing bad managers into four groups.

The Emotional Buffoon: The manager who lacks behavioral competencies. The manager is nonempathetic or, in some rare cases, too empathetic, which can take away from team objectives by catering to the emotional flamboyancy of some team members. Most often, though, this manager is unable to understand or listen to subordinates' needs or wishes. This manager can be too emotional or quick-tempered and lack rationality or emotional intelligence. Sometimes the Emotional Buffoon is thought to be tough or demanding, yet may achieve his or her power through coercion. Subordinates show little if any dedication or loyalty to this socially inept manager.

The Scared Burnout: This poor beast shows little or no initiative. He or she lacks courage to be vulnerable or tough. This manager cannot or will not speak his or her mind or stand by his or her convictions because time and energy has passed them by. The Scared Burnout is the tired workhorse who spent too many years in the field, and yet the thought of dropping out of his or her management position would destroy him or her. In many cases, the Scared Burnout was one of the very best at what he or she did. Years of management abuse and a constant high level of performance pursuit have taken their toll. The Scared Burnout would best serve the company as an individual contributor.

There is a balance that must be struck, and a manager has to understand his or her subordinates while producing a cohesive plan for success. One production manager I spoke to, when asked why he thought some managers failed, told me the following:

> Managers are control freaks. No doubt about that. It's how they got where they are. At that point, they will struggle mightily unless they learn to perform a 180-degree turn. Delegation means you're not going to do it ... someone else is. Delegation requires courage. Managers can destroy months of their own efforts in not trying to get workers to take initiative, take risks and use imagination to reach for the brass ring when they around someone's shoulders, grab the wheel and make a course correction. Quite frankly, they would have been much better off invoking the courage to keep their hands off the wheel, and retaining the worker's loyalty and energy. As a manager, I have never done so well as when I stayed out of peoples' way, and got them the resources they need. I keep my hands and feet away from the moving machinery. And I appreciate it very much when my boss does the same.

In another interview, a line worker said:

> The greatest failures I saw in bosses occurred when they did not give their workers credit for knowing how to do their jobs. I had to laugh at a couple of occasions in my career. Twice, upper management brought in consultants to look at the way we did things and re-invent us. Basically, what the consultants did was what upper management should have been doing: They talked to the production, distribution and sales and office people and made recommendations along the lines of what the people who were doing the jobs suggested!

According to Helanie Scott, President of Align-4-Profit, the keys to management success are "being sincerely interested in what it takes to engage your staff in a way that gets them to take full accountability and ownership. To be a successful manager, or leader for that matter, you must be able to be flexible and put your ego in the passenger seat" (personal communication, August 2, 2008). Scott has established a business built on the shoulders of failure. The paradox is that although people rarely change, businesses can and do.

Method Failures—Failure of Business

There may not be anything more difficult than working on spelling words with your 8-year-old son when he would much rather be outside fishing, playing basketball, or skateboarding. Most every parent has experienced this frustration, and some give in to their child's repeated demands to leave the books behind for a breath of fresh air. Exercise is good for a child, no argument there. Racing around to the point of near exhaustion will not get the child any closer to being able to spell 20 words containing three or more syllables. Why is that? The obvious answer may not be the correct answer. Being outside provides a period of freedom. A field or woods provides countless opportunities for self-expression as well as a cathartic release of built-up frustration and anxiety. So why can't learning how to spell provide the same joy? The answer is found in the way in which the task is introduced as well as the method of teaching and learning. A successful teacher and student will explore various ways to make learning an activity of enjoyment, self-expression, and creativity. Unfortunately, there are many teachers and students who do not approach teaching and learning this way, and the end result is frustration, anxiety, and even failure. One can easily draw a parallel with business.

In business, there have been dozens of reasons why companies fail, including overexpansion, poor capital structure, overspending, bad business location, poor execution with an inadequate business plan, failure to change, ineffective marketing, underestimating the competition, holding on too long to an idea, believing your own marketing spin, and keeping nonperformers. The list could go on and on.

As a young lad, I remember going to the S.S. Kresge Company (we used to call it Kresgees) to buy Snicker Bars for 15 cents or to sit at the lunch counter with my father and munch on bacon, lettuce, and tomato sandwiches (BLTs). The store offered airplanes, inexpensive jeans, an endless supply of goldfish, and the ever-exciting Blue Light Special. Years later, the store changed its name to Kmart and then shortly went out of business.

I recently met Bill Hannon who worked for the S.S. Kresge Company for 22 years prior to its demise. For the past decade, Bill has worked for an industrial chemical company as a sales and service representative spending countless hours in various manufacturing facilities. According to Hannon, the comparisons between Kmart's failure and the various problems that certain U.S. manufacturing facilities experience are similar. Most MBA textbooks will recite all sorts of financial reasons, yet many never seem to address the business dynamic that occurs on a daily basis. The following are different reasons for business failure:

Uncontrolled Control: Upper management takes away control of the floor from the managers. Often a manufacturing line is unique to a particular plant. In a manufacturing environment, the floor manager, plant engineer, or line manager has oversight and authority to make production decisions based on experience and knowledge. A plant that adopts a one-size-fits-all mentality and takes away a floor manager's power to make critical decisions may be doomed to fail. In the case of Kmart, they discontinued autonomy at the local level by taking away the buying decisions, display choices, and local advertisements. The company centralized and systemized everything, which stifled creativity. Local store managers at one time had authorized price adjustment (APA). The store manager could increase or decrease various prices in order to increase sales and profits. This power was taken away from the manager. The store managers had key insight into how and why their customers bought what they did, and they adjusted prices to suit the economic climate. The corporate headquarters did not understand the customers the way the store managers did, and sales suffered.

Cut Costs/Chop Heads: Reduce compensation and eliminate your experience base and you will eliminate your business. The equipment in a manufacturing plant is often customized in order to maximize throughput while reducing downtime. When a company eliminates an employee who spent several years understanding the various intricacies of a machine or a business, the company is sure to suffer. In 1987, Kmart purged the district and regional managers. According to Hannon, "They canned people that were over 40 years old. The employees that built the company with over 20 years experience were replaced with recent college graduates with salaries and impossible over-rides." Kmart also eliminated stock options. At one time, managers were able to apply up to 20% of their income to stock options. Department heads could apply

10% of their income, and stock clerks and register personnel could apply 5% to purchase stock at 50% off the buy price. The company eliminated fiscal incentives such as commissions, bonuses, and "spiffs." Financial reductions have always proved to be a major factor in reduced employee morale. The employees felt they were overworked and underpaid. Employees were asked to unload trucks, set up displays, and cover registers without proper compensation. Kmart also eliminated the departments that were successful but did not have high profit margins. When sales would slump, the company would enforce "cost control," which meant reducing payroll. Another interesting phenomenon happened. The company began to experience product loss not due to shoplifters but rather to employee theft.

Take Away the Soul: Break a company's spirit and break the company. A facility or business has a unique spirit. When that spirit or business soul is compromised, no acquisition will save it. Take a walk into any excellent manufacturing facility, and you can feel the spirit. From the sight of clean tools to the lack of scrap, the plant produces and succeeds. The line workers' faces reflect pride. The plant hums with a certain tempo. Kmart's original Blue Light Special established a carnival-like atmosphere. Employees who announced the Blue Light Special were trained to act like carnival barkers. In a low voice, they would begin "Shoppers, for the next few minutes," and then they would yell, "Are you listening! For the next few minutes we have …." They were taught to romance the product and establish a sense of urgency. Ironically, the Blue Light Special was a flashing red light on a cart. The Blue Light Special was discontinued for several years. Recently, it was reinstated with a picture of a blue light. There was no excitement. The special was not special. When Kmart tried to expand into other markets with stores such as The Sports Authority, Waldenbooks, and Builders Square, it was met with failure. The Sports Authority (acquired in 1990) and Waldenbooks (acquired in 1994) were sold off in 1995. Builders Square stores were closed in 1999. None of the store ventures proved successful. There are as many theories as opinions why the acquisitions failed. Wall Street analysts believed that Kmart was guilty of a failed board that could not provide leadership and a management team that could never settle on a strategic course long enough for it to take hold. Kmart lost its spirit and soul.

Cave Dwellers: Inability to use technology. When data are stored and properly analyzed, data are exceptionally powerful. Manufacturing facilities that have a strong understanding of raw material inventory, quality control, process downtime, and production entitlement will succeed. Information is power. Wal-Mart's information system is legendary. From inventory to procurement to customer purchases, Wal-Mart understands its business climate better than any company does. Wal-Mart also pressures suppliers to reduce prices, and the vendors comply due to the large volumes purchased. Wal-Mart stepped up with information technology and aggressive sourcing while

Kmart stalled. Kmart inventories were reduced due to lack of expertise and a faulty inventory system.

On January 22, 2002, Kmart filed for Chapter 11 bankruptcy protection. Kmart closed more than 300 stores in the United States and laid off 34,000 workers as part of a restructuring. On May 6, 2003, Kmart officially emerged from bankruptcy protection. On November 17, 2004, Kmart announced its intention to purchase Sears, Roebuck and Company. Kmart may be on a slow road to recovery but will never be the powerhouse of Wal-Mart or even Target. Manufacturing facilities can learn valuable lessons from the mistakes that Kmart made.

Failure of Materials, Failure of Machines

Purchasing has little, if any, influence on character flaws or management mistakes, but it can have an influence on the bottom line by understanding the failure of materials and machines. In R. Keith Mobley's *Root Cause Failure Analysis*, Mobley writes, "many of the chronic problems that plague plants are a direct result of vendors' deviations from procurement specifications." An argument can be made that management is responsible for establishing vendor audits and holding them accountable for deviations in performance. Of course, the purchasing specifications can be written to encompass an audit as well as properly define the quality and performance assurance of various items.

For a performance purchasing specification to be developed, it is important to understand the basics of the various failure modes of material and machines. There are basic reasons why failure occurs. Once these reasons are understood, performance parameters can be established, and a specification can be developed. Four sources of failure—*man*, *method*, *machine*, and *material*—are defined earlier in this chapter.

From these four sources spring forth the manifestations of failure. There are hundreds of reference books dedicated to failure analysis. Many of these texts are rich with equations, explanations, and examples of hundreds of failure modes. After reading through a few dozen of these, it became very obvious that failure can be broken down into basic elements regardless of the item. Every failure is a combination of a change and an influence as well as a cadence, articulation, and affect. Thousands of combinations can be the culprit that makes failure analysis an almost impossible effort unless you were to break it down to basic elements. Various items fail for different reasons. When something fails or breaks, it is doing so because of the following physical *change*:

- Deformation
- Fracture
- Wear
- Molecular transition

Change occurs because of the following environmental *influence*:

- Force
- Temperature
- Time
- Chemical

These *influences* have a *cadence*:

- Steady
- Random
- Cyclic

The *cadence* can have variations of *articulation*:

- Amplitude (strong to weak)
- Frequency (fast to slow)

All of these *affect*:

- Surface and work inward
- Inside and work outward
- Entire part, material, device, or tool all at once

Failure can have one or more primary changes and influences as well as one or more secondary changes and influences. Failure can be a combination of several influences that alone would not facilitate change but in concert would bring about a manifestation of failure. Describing the failure mode considers all of these. If a part fails on the surface due to a slow, steady, chemical molecular transformation, it may have experienced surface rust. That is a rather long-winded description for something that is common and understood with two words, but the taximetrics of failure are actually very important in understanding, categorizing, and potentially eliminating the opportunity for failure. By understanding why something fails, you can make a better decision of what to buy. A performance purchase will address each failure mode, look to eliminate the failure or drastically reduce the opportunity for failure, thus driving increased reliability and reduced operational and repair costs.

When we think in terms of a material's strength, we are really trying to understand and define how it will behave in resistance to various forces. The mechanical properties of a tool, material, or device describe how well it will react to physical forces. Mechanical properties occur as a result of the physical properties inherent to each material. These are determined through a series of standardized tests developed by various reputable organizations, such as International Organization for Standardization (ISO), American Society for Testing and Materials (ASTM), American National Standards Institute (ANSI), Society of Automotive Engineers (SAE), and the American Petroleum Institute (API) to name a few. The force is the "push" or "pull" of an object resulting from the object's relationship with another object, be it in a gas, liquid, or solid form. Forces exist as a result of a given relationship to another object.

To understand failure due to force, it is essential that force be understood. There are two types of forces: those that actually involve two or more objects physically touching are *contact forces*, and forces involving two or more objects interacting yet not actually physically touching are called *action-at-a-distance forces*. Examples of *contact forces* include frictional forces, tensional forces, normal forces, air resistance forces, and applied forces. Examples of *action-at-a-distance* forces include gravitational forces, electricity, electrical static dissipation (ESD), or magnetic attraction or repulsion. One would think that pneumatic or air resistance force would be an example of *action-at-a-distance forces*, but air is considered a gaseous object and therefore is a *contact force*. Consider *contact force* as the actual molecules coming into contact with each other, while *action-at-a-distance forces* involve the interplay of subatomic particles.

In general, *force* is defined as a quantity measured using the standard metric unit known as the *Newton* named after Sir Isaac Newton. A Newton is abbreviated by an "N." To say "30 N" means 30 Newtons of force. One Newton is the amount of force required to give a 1-kg mass an acceleration of 1 m/s/s as defined by the following equation:

$$1 Newton = 1 kg \frac{m}{s^2} \tag{3.1}$$

A force is a vector quantity that has both magnitude and direction. The force acting upon an object has both the magnitude (size or numerical value) and the direction. By stating that there were 23 Newtons applied to an object is not correct, rather 23 Newtons downward is a correct description of the force. Because a force is a vector that has direction, it is common to represent forces using diagrams in which a force is represented by an arrow. Keep in mind that forces are vectors; therefore, the effect of an individual force upon an object is often canceled by the effect of another force working in an equal and opposite direction. The effect of a 23-Newton upward force acting upon a steel beam is canceled by the effect of a 23-Newton downward force acting upon the beam. The forces balance out. Of course, there is something to be said for the compression that the beam will experience.

Types of Forces

Applied Force: The force applied to an object by a person or another object. If a machine is pushing a crate across the warehouse, then there is applied force acting upon the crate. The applied force is the force exerted on the crate by the machine.

Gravitational Force: The force by which the earth, moon, or other large object in the universe attracts another object toward itself. It is still unclear how this is achieved. Theories abound concerning how gravity works, yet nothing has been agreed upon yet. *Gravitational force* is considered the weight of the object. All celestial objects have a force of gravity directed "downward" toward their centers.

Normal Force: Support force exerted on an object that is in contact with another stable object. If a box is resting on a floor, then the floor is exerting an upward force upon the box in order to support the weight of the box.

Spring Force: Force exerted by a stretched or compressed spring. An object that compresses or stretches a spring is always acted upon by a force that restores the object to its rest or equilibrium position. For most springs, the magnitude of the force is directly proportional to the amount of stretch or compression of the spring.

Frictional Force: A resistive force exerted by a surface as something moves over it. Two types of frictional forces exist: sliding and static friction. Friction occurs when two surfaces are pressed together. The result is wear and heat. Often, if there is enough localized pressure applied, temperatures in excess of 2000°F can be quickly experienced. This is often the case with seized bearings. The bearings would appear as if someone had melted them with a torch, when, in fact, the act of frictional forces generated the localized temperatures. *Air resistance* is a type of frictional force. The force of air resistance is often observed to oppose the motion of an object.

Tension Force: Force directed along the length of the wire, string, rope, cable, or chain which *pulls* equally on the objects on the opposite ends of the wire, string, rope, cable, or chain.

Strength

Strength has many definitions depending on the type of material and where it is being used. A material's strength is based on a measured ability to perform in a given set of circumstances. A common contention is the World's Strongest Man competition. Competitors will compete in a wide variety of events that test their strength in a host of different venues, such as carrying a large appliance a certain distance as quickly as possible or placing a series of extremely heavy concrete balls on top of pillars. It can be argued (and often is) that some of these tests are not true representations of strength, while events hosted by the Olympics for weightlifting are better indicators of who truly is the World's Strongest Man. The definition is never truly understood, most likely for ratings purposes.

When thinking in terms of strength, it is important to understand. The *strain* is usually expressed as elongation percentage divided by the original length. Most

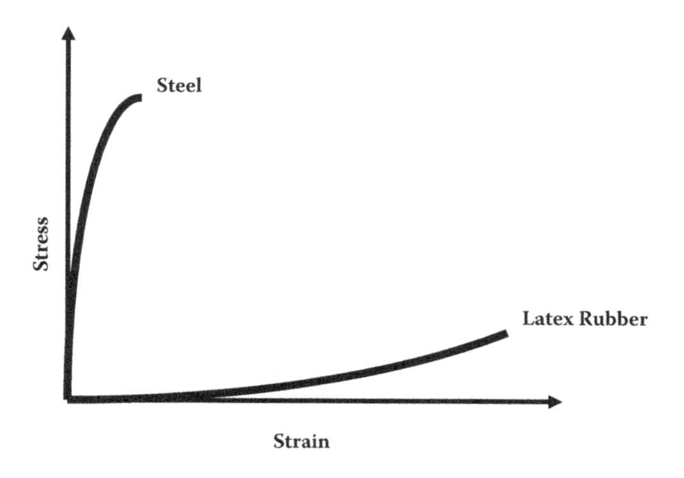

Graph 3.1 Stress-to-strain curve.

materials exhibit elongation prior to fracture or rupture. Even concrete and steel will provide elongation, albeit at a small percentage compared to various plastics.

The *stress* is the *force* causing *strain*. The unit of force (Newton) is usually divided by the cross-sectional area of the material. The Newton per square meter is also called the Pascal, the fundamental unit of pressure, or in this case, of tension (Graph 3.1).

Materials such as steel can resist high stress while producing very little elongation; conversely, latex rubber can be stretched many times its original length prior to failure. The slope of the stress–strain curve is the *elasticity of modulus* (E-modulus) also known as *Young's modulus*. The E-modulus is the ratio of stress to strain defined as:

$$E = \frac{Stress}{Strain} \tag{3.2}$$

Defining Stress and Strain

The *stress* S is the force per unit area as $S = F/A$. The *strain* e is the fractional change in the length as $e = \Delta L/L_0$, where F is the applied force, A is the cross section of the test piece, L_0 is the initial length, and ΔL is the change in length due to stress.

Within the linear, elastic region of the stress–strain diagram, Young's modulus E is defined as the ratio of stress to strain:

$$E = \frac{Stress}{Strain} = \frac{S}{e} = \frac{F/A}{\Delta L/L_0} \tag{3.3}$$

Since the strain is a unit-less ratio, Young's modulus has the same units as the stress, which is Newton per square meter (N/m^2), or Pascal (Pa). In practice, it is usually stated in units of MPa (10^6 Pa) or GPa (10^9 Pa).

It is important to understand the definition for a given material as well as the various tests and units of measure that define strength. This helps in understanding the ability of a material to withstand load or resist cracking or rupturing under pressure or impact. When designing for strength, material class and mode of loading are important considerations. For most plastics, the most common form of strength would be the failure strength or the stress at the point where the stress–strain curve becomes nonlinear. Strength for ceramics, however, is more difficult to define. Failure in ceramics is highly dependent on the mode of loading. The typical failure strength in compression 15 times the failure strength in tension. The more commonly reported value is the compressive failure strength of ceramics. Often adhesives can have adhesive strength and cohesive strength and can be tested according to the direction of the force applied, a peeling direction or a shearing direction, as well as to the various substrates to which they are attached. Paints and coatings are considered strong if they can resist ultraviolet (UV) degradation or are abrasion resistant. A lubricating oil or grease is considered strong if it can continue to provide a boundary film under high loads and resist oxidation. Further details of particular physical characteristics will be covered later in this chapter. For metals, the most common measure of strength is the yield strength. Yield strength is the minimum stress that produces permanent deformation. The yield strength is usually defined at a specific amount of strain, or offset, which may vary by material or specification. The offset is the amount that the stress–strain curve deviates from the linear stress–strain relationship line (Graph 3.2).

Other forms of stress include direct tension, direct compression, bending, shear, direct shear, and torsion. Any further discussion or examination of stress would go beyond the scope of this book and frankly would just stress the reader to the point of yielding.

The yield strength or yield point of a material is the stress at which a material begins to *deform*. Prior to the yield point, the material will deform in an elastic

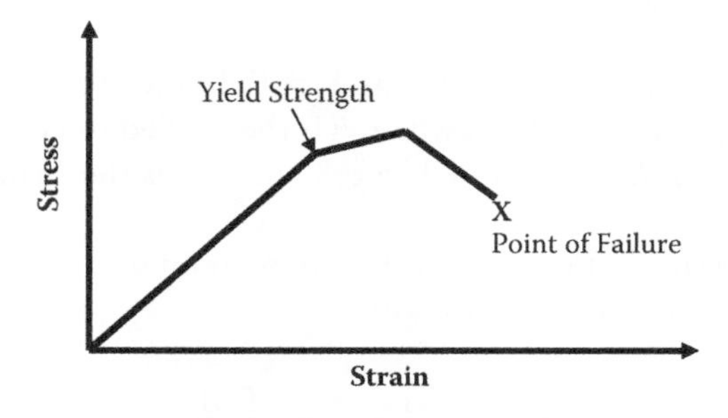

Graph 3.2 Yield strength and point of failure.

fashion and will return to its original shape when the applied stress is removed. Once the yield point is passed, some fraction of the deformation will be permanent and nonreversible. Knowledge of the yield point is very important when choosing material, because it generally represents an upper limit to the load that can be applied. Materials have a variety of stress–strain curves, and there are many different ways to define yielding:

True Elastic Limit: The lowest stress at which dislocations move. This definition is rarely used, because dislocations move at very low stresses, and detecting such movement is very difficult.

Proportionality Limit: The point at which the stress–strain curve becomes nonlinear.

Elastic Limit: Beyond the elastic limit, permanent deformation will occur. The lowest stress at which permanent deformation can be measured.

Offset Yield Point (yield strength or proof stress): Most widely used strength measure of metals, this method allows for a consistent comparison of materials.

Upper Yield Point and Lower Yield Point: The material response is linear up until the upper yield point, but the lower yield point is used in structural engineering as a conservative value.

Toughness: Related to the total area under its stress–strain curve. A comparison of the relative magnitudes of the yield strength, ultimate tensile strength, and percent elongation of different materials will give an indication of relative toughness. Materials with high yield strength and high ductility have high toughness.

Creep (Deformation)

The propensity of a material to slowly move or deform permanently to relieve stress is called *creep*. It is the permanent deformation from prolonged stress exceeding the limit of recovery. Creep is influenced by the magnitude of the load, the time the load is applied, and the temperature. Creep is also influenced by the cycling of the load. Creep is not always considered a failure mode, but it is rather a deformation mechanism. Moderate creep in concrete is sometimes a desired feature with the benefit of relieving tensile stresses that may produce cracking.

Testing consists of applying a load to a test specimen and measuring the strain after a specified time. The dimensions of the test piece are taken into consideration when the test is performed. The *tensile creep* test measures the elongation, breaking point, and strain produced by a load for a predetermined time with constant load, increasing load, or cycled load. The test can be run in a temperature-controlled chamber in order to understand long-term exposure to very low or elevated temperatures. A test sample is prepared by cutting out a sample in the shape of a dog bone. The specimen is clamped into a test fixture, and load is applied separating the

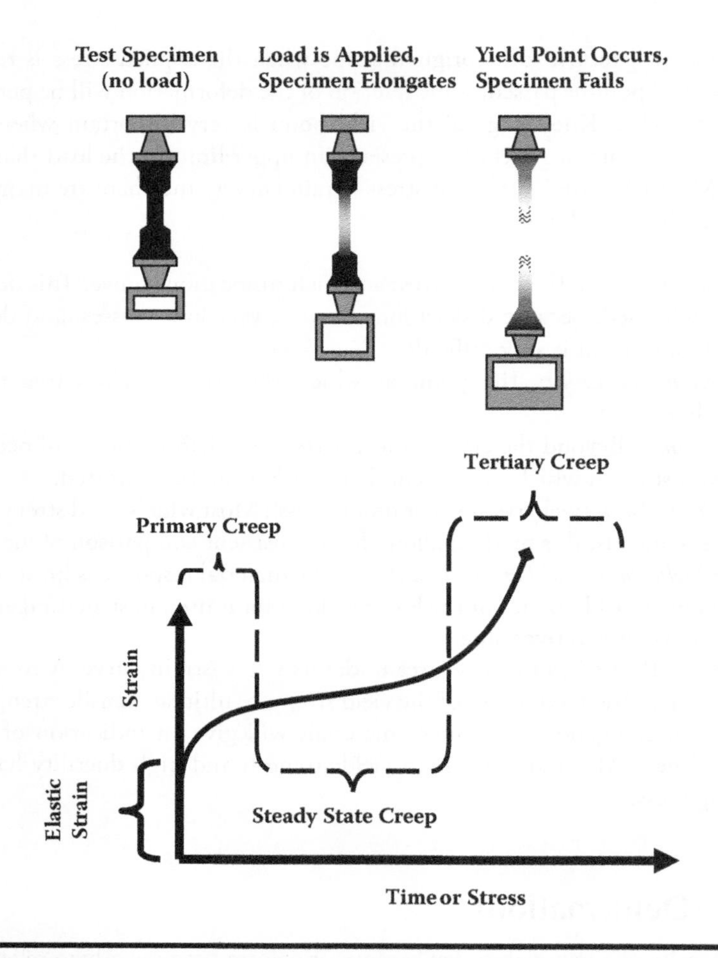

Figure 3.1 Strain versus time or stress creep behavior.

two ends of the test piece. The elongation and tensile strength of a material can be calculated with this method as well as the E-modulus, which is a function of stress divided by strain. This test can be accommodated for bolts, screws, and rivets. A variation of this test is used to understand the adhesive and cohesive strength of various glues, sealants, and coatings.

Creep can be subdivided as *primary*, *tertiary*, and *steady state* creep. The qualitative behavior of strain versus time can be seen in Figure 3.1.

Primary Creep: The first region of Figure 3.1 is the *primary creep* region. The *primary creep* strain is usually less than 1% of the sum of the elastic, steady state, and primary strains. The primary creep region is strongly dependent on the history of the material. If the material had been cycled before the creep test, there would have been many more dislocations present, and the characteristics of the *primary creep* region would have been much different. The dislocations of a material are essentially the loci of the opportunity for failure.

Steady State Creep: The second region of Figure 3.1 is the steady state region. This region is so named because the strain rate is constant. In this region, the rate of strain hardening by dislocations is balanced by the rate of recovery.

Tertiary Creep: When the amount of strain is high, fracture or rupture will occur. In the tertiary region, the high strains will start to cause necking in the material, which can be seen as a thinning of the material. Interestingly, many materials will turn white upon necking, which is a direct response to the aligning of the long-chain molecules. This necking will cause an increase in the local stress of the component, which further accelerates the strain. It is normally a conservative estimate to approximate the end of serviceable life of any material to be the end of the *steady state* creep region because of the short duration of the tertiary region.

Increasing temperature influences the modulus of a material as well as the elastic strain region and the point of failure (Graph 3.3). The temperature range in which creep deformation may occur differs in various materials. For example, tungsten requires a temperature in the thousands of degrees before creep deformation can occur, while ice formations such as the Antarctic ice cap will creep in freezing temperatures. Generally, the minimum temperature required for creep deformation to occur is 30% of the melting point for metals and 40% to 50% of the melting point for ceramics. Plastics have a wide range due to variations. All materials (plastics,

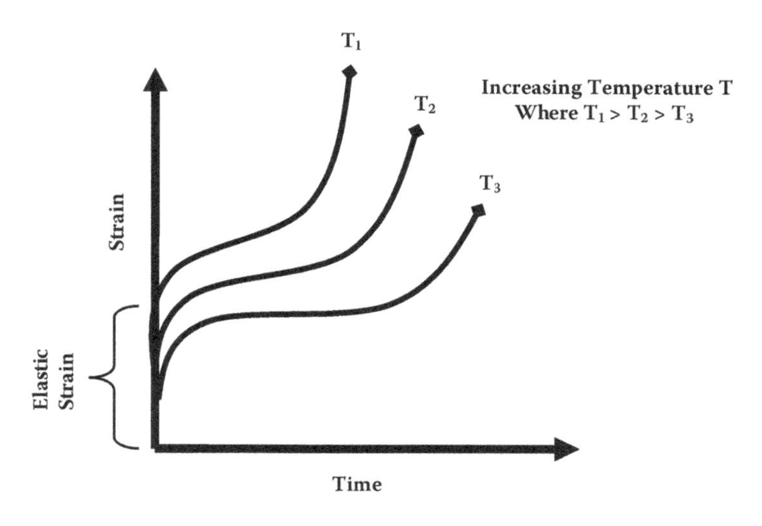

Graph 3.3 Strain versus time creep behavior effect of temperature.

metals, ceramics) will creep when close to their melting temperature. Because the minimum temperature is relative to melting point, creep can be seen at relatively low temperatures for some materials. Plastics and low-melting-temperature metals such as mercury or indium and even lead solder will creep at room temperature.

Fatigue (Material)

Fatigue is the progressive damage that occurs when a material is subjected to cyclic loading. The maximum stress applied is less than the stress limit of the material, yet the material is subjected to a constant or random elongation, bending, or compression that will eventually take its toll and produce a failure.

Typically, a fractured piece will have a dark area that is an indication of slow crack growth, and a bright area that is an indication of sudden fracture (Figure 3.2). Fatigue begins with the crystalline structure of a metal or the long molecules of a plastic or a coating experiencing dislocation. Eventually, small microfissures develop that begin to compromise the physical integrity of the part, tool, or material. Material will not normally recover if fatigue is relieved.

Rate of Fatigue

Temperature, chemicals, and moisture can influence the rate of fatigue. Another consideration that affects fatigue would be the geometry of the part, component, or tool. The quality of the surface has an influence on fatigue as well. Surface roughness, as well as scratches, dings, or dents cause stress or can provide a crack nucleation site that can lower fatigue life. In certain examples, surface stress can be used to increase fatigue life. Shot peening is widely used to put the surface in a state of compressive stress. This will inhibit surface crack formation and improve fatigue life. This technique for producing surface stress is often referred to as *peening*. This improvement is typically observed only in a high-cycle fatigue environment. There

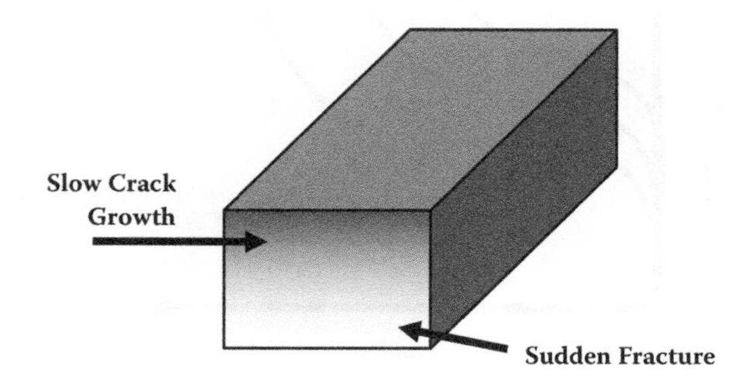

Figure 3.2 Fracture profile.

is little improvement in low-cycle fatigue situations. Another contributing factor would be uneven cooling that leads to a heterogeneous distribution of material properties such as hardness and ductility. Uneven cooling of castings can produce high levels of tensile residual stress that will bring about crack propagation. With that in mind, the size, frequency, and location of internal defects can also play a role. Casting defects such as pores and voids can facilitate fatigue.

A famous example of this type of failure mode was the Liberty ships built during World War II. The Liberty ship was built to replace those destroyed by Germany. Many of the ships developed severe cracks on the hull and deck. This was first thought to be the result of inexperienced welding but later proved to be the metal being used. The ships' metal was exposed to severely cold temperatures that made the metal brittle. Once discovered, corrections were made in order to use the proper metal, and the rest is history.

Fracture

Brittle Fracture

Brittle fracture occurs without any sign of deformation prior to the fracture. Often, brittle fracture affects crystalline materials such as gemstones, glass, many metals, and to a lesser extent, certain types of plastics. Brittle fracture also is dependent on temperature. At very low temperatures, materials that would normally deform prior to fracture at higher temperatures experience brittle fracture at lower temperatures. In Freshman Physics, you may remember the instructor dipping a rose into liquid nitrogen and then dropping it on the floor. The rose most likely shattered into many pieces as if it was made of glass.

Sometimes brittle fracture is not a bad thing. If it were not for this type of fracture, the Stone Age would not have occurred. Early humanoids were able to take advantage of the brittle nature of various stones to create sharp implements that allowed for successful hunting and cutting. Gemstones are "cut" using their inherent brittleness, and even silicon wafers used to manufacture integrated electronic chips are made by cleaving a solid silicon ingot. These materials break because of the latticework of the material's molecules. The molecular structure (also called the crystal lattice) is highly ordered, allowing for a clean break to occur. The smallest possible part of the crystal lattice is called a primitive unit cell. Figure 3.3 presents some examples of typical crystal lattice structures.

In Figure 3.3, the gray sphere in the Cubic Body Centered example is used to signify an internal atom, whereas the white spheres in the Cubic Face Centered and Hexagonal examples signify surface atoms.

Please note that molecules do not actually have sticks holding the atoms in place. The lines are used to help visualize the arrangements. A metal's crystal structure and properties are determined by metallic bonding that is the force holding

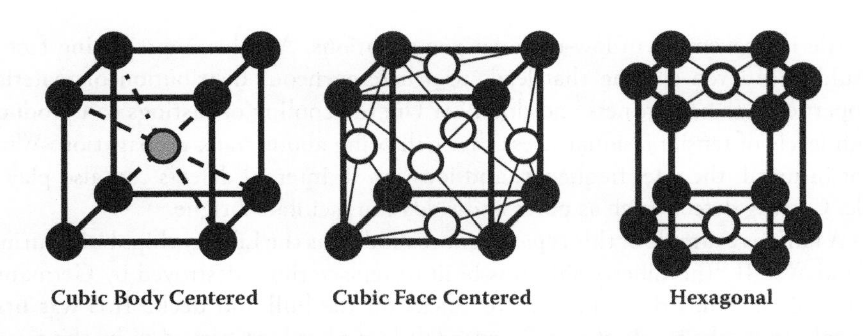

Figure 3.3 Typical crystal lattice structures.

together the atoms of the metal. Each of the atoms of the metal contributes its valence electrons to the crystal lattice, forming an electron cloud that surrounds positive metal ions. These free, negatively charged electrons belong to the whole metal crystal. The ability of the valence-free electrons to travel throughout the solid explains both the high electrical conductivity and thermal conductivity of metals. Solids that do not have an ordered pattern are considered amorphous. Many plastics are amorphous, as is glass. Amorphous solids will experience elongation or deformation prior to breaking. The fracture propagation in an amorphous solid typically does not have a simple path to follow compared to a crystalline structure.

Ductile Fracture

To understand *ductile fracturing*, it is important to first understand what it means for a material to be ductile. Ductility is commonly defined as a material's ability to be deformed without the incidence of failure. Ductility may also be a material's ability to be considered bendable or even crushable. In *ductile fractures*, a considerable amount of deformation takes place before failure. The point at which the failure occurs is controlled by the purity of the materials. At room temperature, pure iron can undergo deformation up to 100% strain before breaking, while cast iron or high-carbon steels can barely sustain 3% of strain.

The basic steps in *ductile fractures* are necking (which results in stress localization at the point on the sample of smallest cross-sectional area), void formation, voids combining to form one large void, the crack ensues, and failure is the result. Figure 3.4 presents a representation of the steps in ductile fracture in a dog-bone sample. As force is applied as the sample necks, voids begin to form and eventually combine. Once the voids combine, the physical integrity of the material is compromised, and a crack in the material forms. Once a crack forms, it is only a matter of time until failure occurs (Figure 3.4).

The heating, cooling, and machine metal can have a drastic impact on the performance of a component. A great example of this is a Japanese samurai sword.

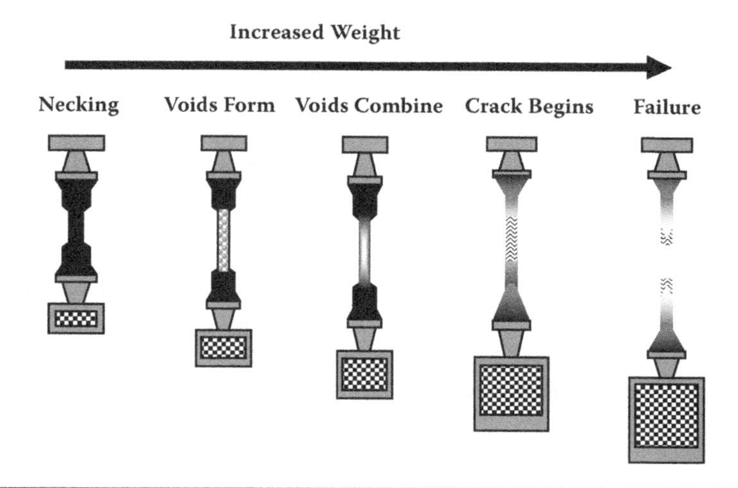

Figure 3.4 Progression of a ductile fracture.

These swords are made by an intricate process of heating the steel, hammering it flat, then folding it, then hammering it flat again, and folding. This process of repeated hammering and folding would be done many, many times. Nick Johnson wrote an interesting article on this process.

> **Interesting Facts on Samurai Sword Manufacture**
> A samurai's sword is his most sacred and prized possession. Not only did the samurai rely on his sword to defend him, but spiritually the sword held greater significance as the samurai actually believed his soul inhabited the sword. Therefore it comes as no surprise that the same discipline and respect in which the samurai wielded his sword, went into the actual making of the sword itself. Swords weren't simply "cast" in a mould and then sharpened. A Japanese samurai sword was made by an intricate process of heating the steel, hammering it flat, then folding it, then hammering it flat again, and folding. This process of repeated hammering and folding would be done up to as much as 30 times, or until the maker was satisfied it had been done properly.
>
> There are quite a few reasons for this labour-intense procedure. Firstly, any air pockets which might develop during the heating of the steel would be eliminated. Having an air pocket in a seemingly solid blade would be a weak point, and any weak point would be seen as neglect and any dedicated artisan would produce the highest quality blades as if his own life depended upon the very blade he was forging. Secondly, in the repeated folding and hammering, what might be

described as "layers" were produced. Take a book and roll it up parallel with the spine, these internal layers would look something like this, almost like the rings of a cross section of a tree trunk. This added much strength to the blade.

Also the natural strengthening carbon elements within the steel, as well as the steel's impurities would be spread throughout the whole of the sword, therefore strengthening it in its entirety. When the blade came to be cooled it wasn't simply quenched in water, another process had to be done first. When steel has been cooled, if it cools from a high temperature right down to cold in a short amount of time, the metal becomes very hard and brittle. Conversely, if steel is cooled slowly from a lower temperature right down to cold, the steel takes on more supple, even softer properties. Because a samurai sword was used primarily as a slicing weapon the blades were subjected to a lot of shock upon impact on the enemy, therefore the blade couldn't be made of the more brittle steel throughout else it would shatter like glass. But the sword had to retain its sharp edge, so it couldn't be made of softly forged steel throughout else it would simply blunt. So a balance was struck using a very clever procedure.

What the Japanese samurai sword makers discovered was by painting on a clay formula onto the blade before quenching, thin amounts onto the cutting edge and thicker amounts onto the back, the steel could be made to take on two completely separate properties, thereby giving the blade the hard cutting edge it required, and the more supple back. Because of the different speeds in which the two halves of the steel cooled this also formed the beginning of the curve from which the sword makers would work to create the famous curved blade.

(Article by Nick Johnson, visit his Web site for more on samurai swords at www.swordsoftheeast.com. Article source: http://EzineArticles.com. Reprinted with permission.)

Wear

Wear is defined as the physical erosion of a solid material surface by the action of another material or force. Tribology is often thought to be the science and study of wear, but actually tribology is the study of surfaces in relative motion. It includes the study of friction, lubrication, and wear. There are four principal wear processes:

Adhesive Wear: Also known as scoring, galling, or seizing. It occurs when two solid surfaces slide over one another under pressure. Surface projections, or asperities, are deformed and eventually welded together by the high local pressure. In bearings as well as in gears, the metal surfaces are actually separated by lubricating oil or grease. If the lubricant fails, the part will fail as well due to the enormous pressures that generate temperatures in excess of 2000°F as witnessed by anyone who has ever had to change a seized bearing. It would have appeared as if the bearings were melted by a torch, yet they were not. As sliding continues, the lubricant became compromised and metal-to-metal contact occurred. The molecular bonds were broken, producing cavities on the surface, projections on the second surface, and, frequently, tiny, abrasive particles, all of which contribute to future wear of surfaces, which contributes to *abrasive wear*.

Abrasive Wear: When material is removed by contact with hard particles, *abrasive wear* occurs. The particles may either be present at the surface of a second material (two-body wear) or may exist as loose particles between two surfaces (three-body wear). Abrasion is the wearing away of surfaces by rubbing, grinding, or other types of friction. It is a scraping or grinding wear that rubs away surfaces. It is usually caused by the scouring action of gritty material such as sand, dirt, slag, and even corrugated cardboard fibers. It usually, but not always, occurs when a hard material is used on a softer material. Consider Babbitt bearings. The metal alloy is soft and would seem to wear out more quickly than a harder metal choice. The structure is actually composed of hard crystals dispersed in a soft metal alloy. As the bearing wears, the harder crystal is exposed, with the matrix eroding somewhat to provide a path for the lubricant between the high spots that provide the actual bearing surface separated by a dynamic film of the lubricant.

Corrosive Wear: Referred to simply as "corrosion," *corrosive wear* is deterioration of properties in a material due to various reactions. Oxidation is a special type of corrosive wear indicated by the flaking off or crumbling of metal surfaces, which takes place when unprotected metal is exposed to a combination of heat, air, and moisture. Rust is an example of iron or iron alloy oxidation. *Corrosive wear* will be covered later in this chapter.

Impact Wear: This is the striking of one object against another. It is a battering, pounding type of wear that breaks, splits, or deforms a surface. It is a slamming contact of metal surfaces with other hard surfaces or objects. A good example of impact wear would be the bucket pins found on construction excavators and backhoes. Often there are only three pins holding the bucket to the arm. As the bucket enters soil or roads, a tremendous amount of force is localized to the pin and bushing surface.

Erosion: The wearing away or destruction of materials by the abrasive action of water, steam, or slurries that carry abrasive materials. Often this type of wear is considered a combination of *abrasive wear* and *impact wear*. Pump parts are subject to this type of wear. Cavitation is considered a type of *erosion wear* that is the result of turbulent flow of liquids that may carry small suspended abrasive particles or aeration that produces microimplosions capable of wearing away metal.

Load can also impact wear, as can direction of the offending surface such as unidirectional sliding, reciprocating, rolling, and impact loads. The speed on the abusive surface as well as temperature, and the angle of contact all affect the wear aspect.

In the results of standard wear tests, the loss of material during wear is expressed in terms of volume. The volume loss gives a truer picture than weight loss, particularly when comparing the wear resistance properties of materials with large differences in density. The working life of components is over when dimensional losses exceed the specified tolerance limits. Wear, along with other failure mechanisms such as fatigue, creep, and fracture toughness, causes progressive degradation of materials with time, leading to failure of material at an advanced age. Under normal operating conditions, the property changes during operation occur in three stages:

Primary Stage: The early stage or break-in period, where the rate of change can be high.

Secondary Stage: The middle age, a steady rate of aging process is maintained. Most of the useful or working life of the component is witnessed at this stage.

Tertiary Stage: The old-age stage, where a high rate of aging leads to rapid failure.

Increased temperature, pressure, strain rate, and stress in the *secondary stage* are shortened, thus reducing the working life. It is important to look to reduce any undue stress and pressures while the device, material, or tool is in the *secondary stage* or middle age. (I will make a point of this to my wife.)

Temperature

Changing temperature of the material affects any reactants and changes the reaction rate. Atoms, molecules, and particles are always in motion. They possess kinetic energy. Temperature is a measure of the material's average kinetic energy. Increasing the temperature will increase the average kinetic energy; therefore, the frequency and energy of collisions will all increase. If the success rate of the collision increases, so does the reaction rate. These conditions will catalyze the reaction and allow for greater mobility of the particles and reactive species. With great mobility comes increased opportunity for a reaction to take place. Conversely, certain conditions or substances will actually retard the reaction process or inhibit it altogether. In this instance, the temperature has a reduced effect. Antioxidants, rust, and corrosion inhibitors work along these mechanisms.

The increase in reaction rate with a minimal temperature change is so dramatic because the overall distribution of kinetic energies in a sample of particles must be considered. There is a significant proportion of particles having a very high kinetic energy and also those with very low kinetic energy. This overall concentration is called a normal distribution curve.

An example of a distribution curve would be the height distribution of American males. The average height of the American male is 5′10″. A certain percentage is above 6-feet tall, an even smaller percentage is above 6′6″ tall. Conversely, a certain percentage is below 5′4″, and still a smaller percentage is below 5 feet. The percentage gets smaller as the height is further away from the average height. This fact is consequential when a college is looking to recruit a center for the basketball team. These outliers are not the norm but rather are isolated examples. There is not a prevailing requirement for 4′2″ 7′4″ people, but their existence can make a very big impact on a basketball team or as a horse jockey. Now imagine being able to skew the average height up or down by a few points. Change the average height to 6′4″. The opportunity now exists for an 8′2″ individual (see Graph 3.4). That could make a very big impact (especially for the New York Knicks). The change in temperature makes the same dramatic impact.

The same can be said for the size and kinetic energy of the reactive components and particulate. If the temperature is raised a little, the average kinetic energy of the particles increases, but the distribution curve size remains the same. The new proportion of particles possessing energy for activation of the reaction is much greater, so the reaction rate increases dramatically. In Graph 3.5, increasing the temperature moves the curve to the right, allowing the maximum amount of molecules to have the needed energy to react.

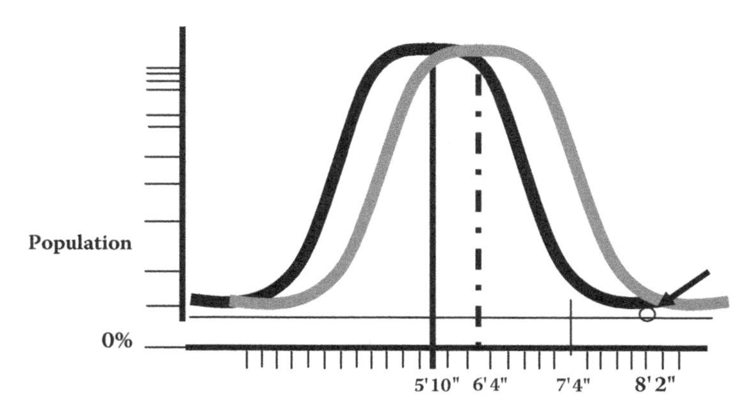

Graph 3.4 Size distribution.

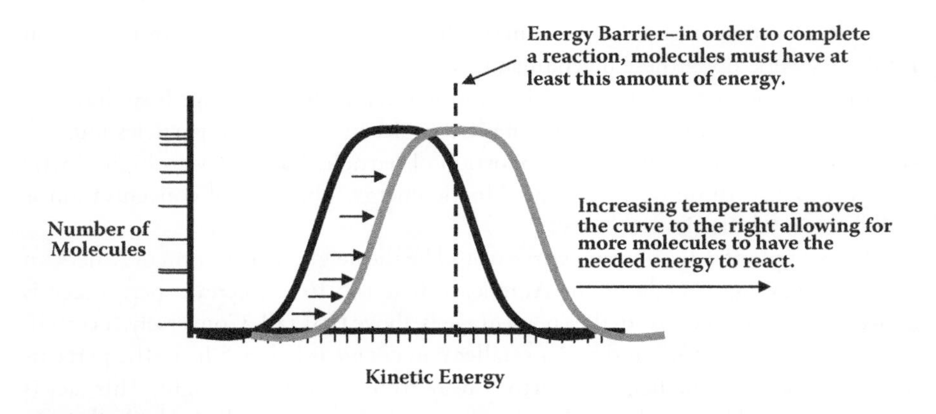

Graph 3.5 Kinetic distribution.

The true kinetics of a reaction can only be determined through experimental methods. Reactions have many variables that can influence the rate. It would be a mistake to assign a rate equation to a given system without further investigation.

Thermal Shock Failure

Thermal shock is the name given to material cracking from rapid temperature change. Glass and ceramics are vulnerable to this failure due to lack of toughness, low thermal conductivity, and high thermal expansion coefficients. Glass and ceramics are used in high-temperature applications because they have a high melting point. The shortcomings can be addressed through processing or design modifications. Toughness or thermal conductivity can be improved upon relatively easily, but changing a material's melting point is far more complicated.

Thermal shock occurs when a thermal gradient causes different parts of an object to expand by different amounts. This differential expansion can be understood in terms of stress or of strain, equivalently. At some point, this stress overcomes the strength of the material, and a crack forms. If nothing stops this crack from propagating through the material, it will cause the object's structure to fail. A good example of this would be the use of cast aluminum as a material choice for an engine block and steel as the material choice for the head and oil pan. Even the best seal would have a difficult time stopping engine oil leaks due to the difference in thermal expansion and contraction rates of the materials. Another example would be how ice cubes will quickly crack when put into a cup of hot tea. The outer surface of the ice cube exposed to the hot tea will expand while the inside remains the same dimension. A crack ensues.

Reinforced carbon-carbon composite is extremely resistant to thermal shock due to graphite's extremely high thermal conductivity and low expansion coefficient.

The high strength of carbon fiber used in the composite also provides an ability to deflect cracks within the structure.

Oxidation—Molecular Transitions and Chemical Influences

Corrosion

Corrosion is typically defined as the oxidation of metal occurring in the presence of moisture. The most common form of corrosion is rust. Rust occurs when iron or an iron alloy is exposed to water, humidity, or steam. The iron reacts with the oxygen that is in water to form rust. The amount of water available to provide the iron with enough oxygen to form rust will also determine the color of rust, which varies from black to yellow to orange-brown.

It is often taught in high school chemistry that oxidation is the loss of electrons, but that is not entirely correct. Electron transfer does not happen. It becomes very complicated to describe the mechanisms at work accurately, so in general terms and for lack of time and patience, oxidation is described as the loss of electrons. The actual chemical process of oxidation involves an increase of the oxidation number, which is the charge it would have if it were stripped of any attached atoms. Often the oxidation number is the same as the oxidation state, but there are a few exceptions. Chemicals that have the ability to oxidize (change the oxidation number) are known as oxidizing agents. The substances remove the electrons from the material. Oxygen, chlorine, and bromine are examples of oxidizing agents.

The presence of oxidation on metal can often be seen on rusted iron parts. A serious problem that often occurs is that the formation of rust occurs away from pitting, which can be hidden. This is possible because the electrons produced during the initial oxidation of iron can be conducted through the metal, and the iron ions can diffuse through the water layer to another point on the metal surface where oxygen is available. This process results in an electrochemical reaction where iron serves as the anode and the oxygen as the cathode, and the water that contains reactive ions acts as a transport medium (Figure 3.5).

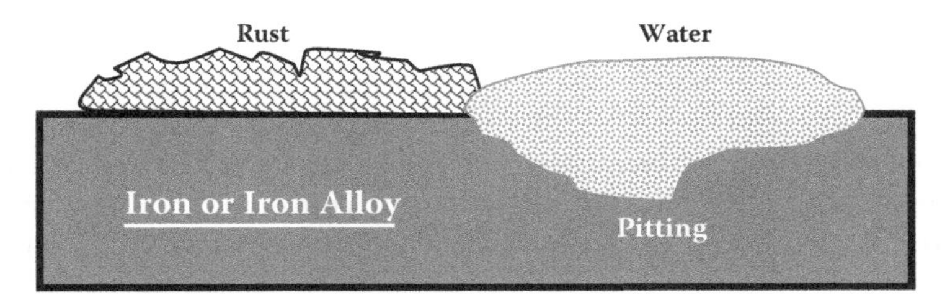

Figure 3.5 Pitting development.

Heightened concentrations of chlorine-containing chemicals can actually interfere with a metal's ability to form a protective oxide layer or passivating film. Very small local fluctuations will degrade the oxide film in a few critical points, and then corrosion intensifies, thus causing pitting. A pinhole on the surface can hide a quarter-size pit just below. These problems are especially dangerous because they are difficult to detect before a part or structure fails.

Many other factors affect the rate of corrosion. Saltwater greatly increases the rate of rust development. This is because saltwater increases the conductivity of the aqueous solution formed at the surface of the metal and enhances the rate of electrochemical corrosion. This is one reason why iron or steel tends to corrode much more quickly when exposed to salt used to melt ice on roads or moist, salty air near the ocean. Rust is a metal oxide that lacks any strength and will not adhere to the surface. This is one reason why pitting occurs. Extensive pitting eventually causes structural weakness and disintegration of the iron or iron alloy. Interestingly, aluminum will form a very tough oxide coating that strongly bonds to the surface of the metal, preventing the surface from further exposure to oxygen and corrosion. Corrosion can occur in acidic or caustic environments and can affect nonmetallic materials, such as plastics and even glass or ceramics.

Corrosion in Nonmetals

Nonmetal materials such as plastics, ceramics, and glass typically do not corrode the same ways that metals do, mainly because they do not conduct electricity or readily change oxidation states. In the true sense, these materials do not corrode, but they can break down in the presence of various chemicals such as acids, caustics, and solvents.

Most corrosion-resistant plastics generally fall into two groups: polyolefins (polyethylene and polypropylene are examples) and polyvinyl chlorides. These types of plastics have excellent chemical resistance and are low cost. The polyolefins have excellent resistance to many solvents but are not chemically bondable. If bonding is required, these types of plastics can be bonded by thermal welding with low-cost equipment. Polyolefins can be produced in sheet, rod, tubing, and film.

Other materials such as fluoropolymers (Teflon® is an example) also exhibit excellent chemical-resistive properties but are very difficult to formulate or process, which results in higher costs. The polyvinyl chloride plastics, including PVC and CPVC (also known as chlorinated polyvinyl chloride), have outstanding chemical resistance properties but can be up to 40% heavier and structurally more rigid than polyolefins. PVC is the oldest corrosion-resistant plastic. The properties of PVC and CPVC are typically identical except that CPVC has better temperature performance. Both PVC and CPVC are chemically and thermally bondable. These plastics can be made into sheet, rods, and tubing, as well as various profiles.

Galvanic Corrosion

Essentially, galvanic corrosion occurs for the same reason the potato can be made into a battery. The battery was invented around 1800 by Alessandro Volta. It is a great example of how chemical energy is converted into electrical energy. There has been some controversy suggesting that simple batteries may have existed a few thousand years ago. This has yet to be accepted by the scientific community or the Italian government. If you ever created a potato battery, you established a galvanic potential that could power a flashlight.

You do not have to use a potato, an orange or apple would work as well. Insert a piece of copper wire (14 gauge works well) and a galvanized nail (preferably a 5 penny or larger) into the potato. (You will have to sand off the oxide layer of the wire and nail prior to tuber insertion.) The nail serves as the negative (–) or cathode of the potato battery, and the copper wire is the positive (+) or anode of the battery. Make sure that the nail and wire are about an inch apart. You can test the voltage of your battery using a multimeter. Normally, a potato battery will not produce more than a few milliamps of current. You have to put the potato batteries in a series in order to have enough energy to power a small lightbulb. The zinc found on the nail gives electrons, and the copper strip accepts the electrons. Because the zinc strip frees electrons and the copper strip uses electrons, if you put a wire between the two strips, then electrons will flow from the zinc to the copper, producing electrical energy. The potato serves as a means of holding the leads in place while allowing the flow of electrons through the water (electrolyte) that is present. In corrosion, the positive (+) anode is consumed or in this case corroded. The current will consume the anode material while the cathode material remains unphased.

Factors such as relative size of the anode (+) material, the type of metals used, and operating conditions such as temperature, humidity, and salinity will have an effect on galvanic corrosion. The surface area ratio of the anode and cathode will directly affect the corrosion rates of the materials. Metals that are high on the galvanic table (the cathodes), such as gold, platinum, and silver, will remain unchanged during the galvanic environment, and the anodes such as calcium, sodium, and aluminum will lose metal and corrode. This is an important consideration if a component has dissimilar metals.

Deposit Formation

Many gearboxes, hydraulic systems, and engines can quickly become contaminated with water, particulate, and various deposits such as varnish and sludge that are the products of oil oxidation. These contaminants contribute to the degradation of the lubricant and increase operating temperature, energy demand, component wear, and oil usage. Even new systems can often contain contaminants that contribute to these problems. There is technology available that will remove contaminants from the systems while in operation and prepare the metal surfaces to readily accept the surface-active agents found in performance products. This technology has been proven to increase equipment life while decreasing wear metals, operating temperature, and energy demands.

To condition a system for improved reliability, a basic understanding of how and why various deposits develop must be understood. Deposits develop when the lubricant breaks down into other compounds. By understanding this process, certain measures can be taken to slow the formation of deposits. Lubricants can break down and create sludge, varnish, and deposits due to several reasons (Table 3.1).

Table 3.1 Sources of Deposit Formation

Contributing Factor	*Mechanism*
Inferior base oil	When the base oil contains one or more of the following: light fractions, sulfur, low molecular weight hydrocarbons
Insufficient concentration of performance additives	Low concentrations of antiwear agents or friction modifiers produce wear metals and localized frictional heat Low concentrations of demulsifiers can lead to emulsified oil, which hastens oil oxidation
Yellow metals (copper, brass, bronze)	Contaminants from external or internal sources can catalyze oil oxidation and lead to deposits and additive depletion
Heat	High ambient temperatures, operational heat (from kinetic energy or combustion), frictional heat (from metal to metal contact), and pressure (from aeration or operation) can accelerate oil oxidation and additive depletion
Water	By-products of combustion or external contaminants can hasten oil oxidation and lead to deposits and additive depletion
Acids	Develops when base oil breaks down into reactive species, produces sludge, varnish, and other deposits, or an external source such as acid washing Common forms are sulfuric acid, nitric acid, and various carboxylic acids
Caustics and solvents	External contamination breaks oil molecules into reactive species; the reactive species may polymerize to form resinous deposits

When lubricants oxidize, they form reactive materials that can reconstitute into different deposits. Varnish, lacquer, sludge, and gum are several of the typical deposits formed and the problems that can occur when lubricating oil breaks down. Varnish, lacquer, gum, and sludge are basically the result of oxidized oil, and are all carbon-based materials. These materials are typically large in molecular size compared to other compounds found in lubricants, save the base oil. These deposits are formed by chain polymerization. There are two methods that are commonly understood to explain the creation of large molecules.

The two methods—step polymerization and chain polymerization—generally synthesize large molecules. Step polymerizations proceed by a stepwise reaction between chemically functional reactant molecules. The size of the molecule increases at a relatively slow rate. One reactive molecule combines with a second; they form a single reactive molecule, which then reacts with a third, and so on until eventually large polymer molecules have been formed. Chain polymerizations require an initiator in the form of a free radical or reactive specie. Free radicals and reactive species (anions or cations) can be generated by many of the conditions found in gearboxes, transmissions, hydraulic systems, or engines. Chain polymerizations occur by the propagation of the reactive molecule by successive additions of large numbers of other reactive molecules. Graph 3.6 shows the typical reaction rate and development time of the two polymerizations.

There are several reasons why deposits form in a lubricating system. Typically, the oil undergoes a transformation from one molecular structure to another. This can be in the form of a mixture of a contaminant, such as fuel soot with motor oil or transmission sludge from debris and gear oil. It can also be varnish or lacquer on a piston pump cylinder from excessive temperature and load. More often, it is

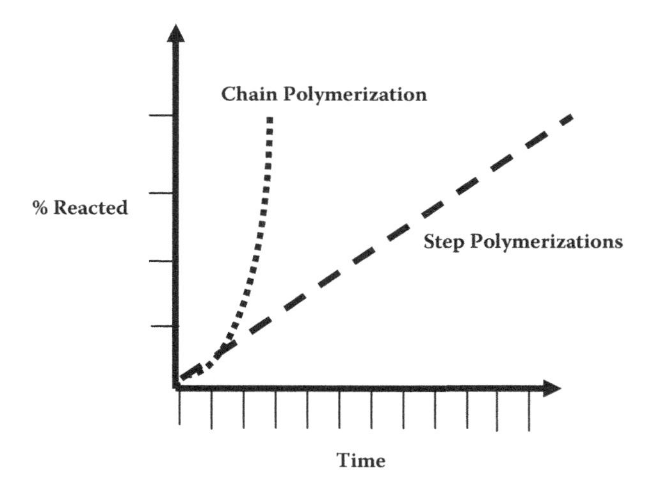

Graph 3.6 Polymerizations.

the breaking down of the base oil and the building up of a new compound. This reconstitution develops through a series of steps. The first step is the generation of a reactive compound or free radicals. There are several ways in which these compounds develop. Once in the system, they combine or polymerize into new compounds in the form of deposits. Various environmental conditions can facilitate the generation, such as temperature, pressure, water, solvents, acids, caustics, and various metals.

Free radicals can be developed by a few different methods. One is through mechanical energy such as milling, pressure, and high pressure combined with shearing deformation, or shock waves through a nonthermal process. Another way is from acids that cleave the molecular bonds. A high concentration of short or moderate chain-length molecules of the oil can break, creating a higher population of free radicals. These short reactive species readily bond with other like species or contaminants and polymerize into deposits. A long chain-length oil molecule will produce less reactive species due to bond length and restricted molecular mobility. Excessive heat and pressure can also break the molecular bond and form free radicals that, in turn, will polymerize into deposits.

There are countless molecular arrangements that make up deposits in any given system. Deposits are typically the result of the base oil breaking down into a reactive compound. Once the reactive compound is developed, it reconstitutes by reacting with another reactive compound forming a deposit. The typical reaction is considered a free radical chain polymerization. The reaction occurs in three steps:

1. *Initiation*: Free radical chain polymerization is a type of polymerization in which the propagating species is a long-chain free radical, usually initiated by the attack of other free radicals, acids, or reactive species that are derived from heat, water, acids, contaminants, and so forth.
2. *Propagation*: The polymerization proceeds by the chain reaction addition of reactive molecules to the free radical ends of growing chain molecules.
3. *Termination*: Finally, two propagating species (growing free radicals) combine (disproportionate) to end the polymerization reaction and form one or more polymer molecules in the form of a deposit (sludge, varnish, lacquer, etc.).

The rate at which the reaction takes place actually accelerates with time (and heat). It would be natural to assume that the reaction rate would slow with time because the concentrations of reactive molecules and initiators reduce as they are reacting. The exact opposite is true. Three routes also known as diffusion-controlled termination steps explain this behavior. The first route is a translational diffusion of two of the propagating radicals until they are in close proximity to each other. The second route is the segmental diffusion of the polymer chains. In this route, the rearrangement of the two chains occurs so that the two radical ends are close enough to react with each other. The third route is the actual chemical reaction of the two radical ends to form the polymer or, in this case, the deposit. During the course of the reaction, the translational diffusion route decreases faster than the increase rate of the segmental diffusion route. This is how rapid autoacceleration occurs.

Due to the speed at which the deposits can develop, it is very important to act quickly when a system shows signs of early contamination. Free radical chain polymerizations can occur very quickly. If the onset of oil oxidation or the free radical polymerization of the reactive species is not quickly addressed, very serious problems can occur.

Factors That Affect Deposit Formation

Concentration and Pressure

When the concentration of the reactive compounds, free radicals, acids, or contaminants is increased, the molecules or particles are closer together. There are more molecules in a given volume, and the same applies when pressure is increased. When this occurs, there will be more collisions or opportunities for a reaction to take place, so there will be a greater chance of successful reactions occurring. The rate of reaction increases. The reactions are the generation of acids (sulfuric, nitric, carboxylic), free radicals, and the polymerization of the free radicals into deposits. In many reactions, it is not convenient to measure concentration. Other properties can be monitored that directly represent the concentration. For instance, a change in chemistry (see Graphs 3.7 and 3.8), loss in mass (density), the production or potential for production of oxygen or another gas (bomb calorimeter), change in color or clarity, change in pH, change in conductivity, or change in pressure. Many of these changes can be tested on site using various methods or using oil analysis laboratories.

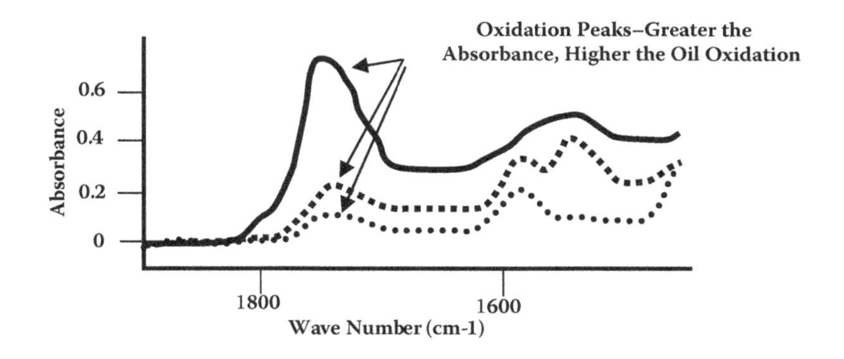

Graph 3.7 Fourier transform infrared (FTIR) scan of oil samples with varying degrees of oxidation.

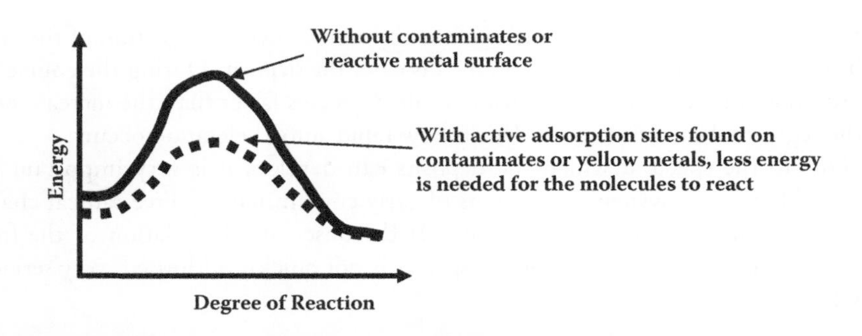

Graph 3.8 Degree of reaction.

Particle Size and Contaminant Type

There are five fundamental types of wear that can produce particles:

- Rubbing
- Cutting
- Rolling
- Fatigue
- Severe sliding
- Combined rolling and sliding

These wear mechanisms were covered earlier but are present in almost all systems. Wear particles can occur in many different systems and are typically the precursors to free radical development and oil oxidation. Microscopic (submicron up to 50-micron) contaminants such as wear metals, dirt, and debris always react faster than lump solids. The small contaminants have a much greater total surface area than lumps, so there will be many more chances for reactions to take place. There will be an increased number of successful collisions, so the reaction rate increases. Only rubbing wear and early rolling fatigue mechanisms generate particles predominantly smaller than 15 microns. These are particularly dangerous. The chemical nature of the contaminant can influence the development of deposits. The contaminant can be a metal ion or particle or organic or inorganic particle or even a fiber from corrugated cardboard. A particulate provides a different reaction mechanism, which has lower activation energy, allowing a greater proportion of all the molecules to collide. Solid contaminants or metal can provide a surface on which the reactant molecules can temporarily stick in the correct orientation for an easy reaction to take place.

There are two ways in which metals and contaminants can facilitate the oxidation process. When two different molecules bump into each other, they might react to make new chemicals. How fast a chemical reaction occurs depends upon how frequently the molecules collide. Metal particulate, inorganic, or organic contaminants make a chemical reaction go faster by increasing the chance of the molecules to collide. The first method is by adsorption; the second method is by the formation of intermediate compounds.

Adsorption

Adsorption occurs when a free radical, reactive species, or reactive oil molecule sticks onto the contaminant or reactive metal surface (typically yellow metals). The following is an example. Copper is a typical metal found in many valves and bearings. It can act as a catalyst when in contact with sulfur. Sulfur is found in many base oils and certain additives and even fuel. Sulfur and oxygen react, and the resulting reaction produces sulfur dioxide and then sulfur trioxide. Sulfur trioxide is very reactive with water and produces sulfuric acid that will corrode metal, break down oil to form reactive species, and can deplete additives. The molecules of the sulfur and oxygen get adsorbed (stuck) onto the surface of copper very easily. Because the two molecules are held so close together, it is more likely that they will collide and therefore react with each other (Figure 3.6).

The sulfur and oxygen have a strong attraction to react. The sulfur dioxide easily falls off the copper surface, leaving space for more sulfur and oxygen to react and form sulfur dioxide and then sulfur trioxide which will then react with water to form sulfuric acid. Less energy is required for the molecules to react (see Graph 3.8). The copper acts as a meeting place for the reactive species to bond. The same can be said for wear metals and outside contaminants.

Some deposits develop by forming intermediate reactive compounds. The chemicals involved in the reaction, in this case, are reactive species such as free radicals or reactive oil molecules combined with the metal or contaminant that make an intermediate compound. This new compound is very unstable. When the intermediate compound breaks down, it releases the new compounds in the form of deposits and more contaminants to further react. The sulfur dioxide example is one of several that occur. Carboxylic acids and nitric acids can also form, but the reactions are very complicated and go beyond the scope of this chapter.

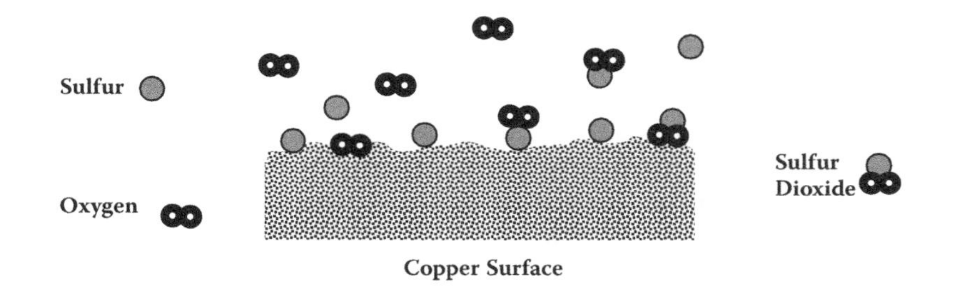

Figure 3.6 Sulfur dioxide development.

Documenting Failure

To accomplish this, it is essential that management spend time on the floor understanding why machines break down. Once this occurs, management must also spend time understanding the legitimate corrective actions and how to successfully implement these changes. In many instances, what looks good on paper may not always make the most sense on the floor. A simple checklist can aid in understanding how failures occur.

- What happened?
  - Define incident and equipment specifics.
  - Interview all personnel directly and indirectly.
  - Gather any physical evidence (worn gear, seized bearing, etc.).
- Where did it happen?
  - Identify specific area affected.
  - Has this occurred before and on different equipment?
- When did it happen?
  - Identify time frame of occurrence.
  - What was the sequence of events that led up to the failure?
- List changes, if any:
  - Products
  - Practices
  - Procedures
  - Environment
- Who was involved?
  - Personnel directly and indirectly involved
  - Supervision and procurement and purchasing
- What was the impact?
  - Was anybody injured?
- What was the financial impact in terms of downtime, parts replacement, labor costs, and direct and indirect cost pools?

Anytime management interviews technicians involved in failure, it is an uncomfortable, if not hostile, situation. Care should be taken to convey the message that the questions asked are required. The questions are asked not to assign blame, but rather to fix the system. The relationship between subordinate and management can be tenuous when compounded with a costly situation such as equipment failure. Production and maintenance technicians may not be straightforward as to the actual reasons why a particular incident occurred. Sometimes the line workers or mechanics may feel as though the reasons for failure are management or company related and bringing them to light may cost them their jobs. In some cases, an outside "interviewer" has to be called in who can develop the background data impartially and in confidence. This provides a useful proposal that will reduce, if not eliminate, the failure that has occurred.

To use this information effectively, the following six steps can be taken to document the failures and set the groundwork for identifying the various standards to promote increased reliability:

1. Develop a failure definition, contact flow diagram, gap analysis, preliminary worksheet, and interview schedule.
2. Interview facility personnel to determine what the failures are, the frequency, and the impact.
3. Input into an electronic spreadsheet and determine any redundancies.
4. Determine the true cost of these failures.
5. Determine "Significant Few"; determine 20% or less of failures that result in 80% of the losses.
6. Verify that the results are valid.

Once you have done this, many of the obvious situations will be validated, but other situations will present themselves that were not normally considered. The use of an equipment maintenance management application will cut hundreds of hours of time required to do this work.

First, we must understand why we need to buy something to begin with, because what we have is broken or is not lasting long enough, or is costing us more than it should. In other words, it has failed us. We must establish a root cause of the failure and then identify the performance requirements and the standards that address those properties to offset the failure. Only then can the performance-based purchasing specification be properly written. This process is only going to make sense for the things that have failed us that end up costing (directly and indirectly) thousands of dollars due to the failure.

If you are looking to buy a case of nails or a bucket of paint, this process is not for you. On the other hand, if you are looking to purchase material or a device that has a direct influence on your operation, then by all means take some time and do the work. The process outlined in the first few pages does not take as much time as you would think, but it does require a systematic approach and decent documentation skills.

In the previous sections, we examined the costs of failure, the sources of failure, and the physical properties that would address these. To develop a comprehensive purchasing specification, it is absolutely essential that you assemble a team of mechanics or technicians as well as operators, if appropriate, and understand the root causes of the failures. This is important even if the part or device is not perceived as actually failing. Remember, the item is being replaced for a reason, and your task is to procure a replacement that will provide better performance that, in turn, will drive down the overall cost of operation.

To get started, you have to establish the root cause of failure. The steps in root cause identification are:

- *Examination*: Of the failed part or material and the system in which it operates. Care must be taken to inspect parts for damage, contamination, moisture, cracks, or other signs of stress.
- *Identify the Failure*: Look for patterns, manifestation, form, or arrangement of the failure.
- *Application*: A close examination of the work performed by the motor and the characteristics of those types of loads, as well as the operating environment and the demands put on the material or device.
- *Maintenance History*: A look at the work performed, and note any replacement parts along the way.
- *Interview*: Ask the folks who are in close contact with the material or part. Remember this is not an inquisition but rather an enquiry.

When done properly, all relevant information pertaining to the application, appearance, and maintenance history would be available prior to the actual inspection of the material or failed part. In real life, however, the process usually consists of inspecting the failed part or material and then acquiring information about the application and its maintenance history. Document the information. It will be used later to help establish the specifications. At the end of this section will be several worksheets that will prove to be simple to use and valuable to have.

During the analysis, information regarding the product or its components is vital. The following information is a good starting point to begin the investigation. The analysis can be carried out for a specific project, a specific product, or a product subsystem or component. An option for you to consider is a severity grading system. The following is an example. Use the bullet points as topic categories when assembling your data. This may be required to help sort through many different items that you are considering writing a specification for. List the following:

System/Component/Function: The specific name of the element for which the analysis is performed is noted.

Potential Failure Modes: The way a product, a product subsystem, or a component can fail while it is being used is noted. The person performing the analysis must at this point be creative and think of all possible ways that the product can be used and fail.

Potential Effects of Failure: For each potential mode of failure, a potential effect is noted. A detailed description is required. A potential mode of failure could have several effects. All must be noted.

Severity: Each failure effect is rated for its severity, typically on a scale of 1 to 10, with 1 representing the least serious and 10 the most serious. Typically, each effect is scaled as follows:

(9–10) Potential safety risks and legal problems, potential loss of life, or major dissatisfaction

(7–8) High potential customer dissatisfaction—potential serious injury or major product mission disruption

(5–6) Medium potential customer dissatisfaction—potential for small injuries, product mission, and inconvenience delays

(3–4) Customer may notice the potential failure and may be a little dissatisfied or annoyed

(1–2) Customer will probably not notice the failure—undetectable failure

Critical: The quick identification of critical failures can be achieved by noting them in the column provided.

Potential Cause of Failure: Each failure must have a cause. It is very important to identify the cause and note it. Possible causes can be wrong tolerances, poor alignment, operator error, component missing, fatigue, defective components, lack of maintenance, and so forth.

Occurrence: It is necessary to note the likelihood of failure occurring. A probability assessment must be made, and scores 1 to 5 must be noted for each failure by the person performing the analysis. A typical scoring board is given below:

(5) Very high probability of occurrence

(4) High probability of occurrence

(3) Moderate probability of occurrence

(2) Low probability of occurrence

(1) Remote probability of occurrence

Current Design Controls: Design controls that can be used to reduce or eliminate the potential failure must be identified and noted.

Detection: A scale must be set for failure detection and the results of the scaling noted. A typical scaling board is given below:

(5) Zero probability of detecting the potential failure cause

(4) Close to zero probability of detecting the potential failure cause

(3) Not likely to detect potential failure cause

(2) Good chance of detecting the potential failure cause

(1) Almost certain to identify potential failure cause

It is likely that more than one failure mode will be identified during the analysis. It is important to prioritize these by assigning a Risk priority No to each of them. This number is calculated according to the following equation:

$$\text{Risk Priority \#} = \text{Severity Rating} \times \text{Occurrence Rating} \times \text{Detection Rating}$$

This number will provide information regarding the design risk, and it is best to give immediate attention to the items with the highest number. This exercise will assist in developing information for the development of a comprehensive purchasing specification.

Summary

At this point, you should have a firm grasp of understanding why a tool, component, or part fails. This will prove essential for the development of a comprehensive purchasing specification. As the failure mode is understood, then the documentation process can begin in earnest. You have now been introduced to various sources of failure. This information is extremely important and valuable when reviewing with the mechanics and technicians as well as vendors who will be pitching solutions. Having a good grasp of failure sources will help establish a very strong foundation for a performance-based purchasing specification.

Tools for the Specification Development Process

The following work sheets can be used to begin the specification development process. The total cost of failure (TCF) work sheet will help establish the significant items that should be considered by way of examining the specifics on the failure mode. Often this exercise helps weed out preconceived ideas that certain items or materials either cost the company more than thought or less than they actually do. The Failure Mode work sheet will then help define the problems you are trying to solve.

First fill out the TCF and then choose what the greatest cost items are. Then fill out the FM to begin the specification process.

Total Cost of Failure (TCF) Work Sheet

TCF Primary Work Sheet						
	A	*B*	*C*	*D*	*E*	*F*
Incident and Reason	*Frequency*	*Item/ Part and Cost*	*Downtime Cost*	*Labor Costs*	*Additional Costs*	*Total Cost of Failure A* (B + C + D + E) = TCF*

The Total Cost Factor (TCF) Detailed Work Sheet (Used as a Supplement If Management Requires More Details)

What happened?
• Define incident and equipment specifics.
• Interview all personnel directly and indirectly.
• Gather any physical evidence (worn gear, seized bearing, etc.).
Where did it happen?
• Identify a specific area affected.
• Has this occurred before and on different equipment?
When did it happen?
• Time frame of occurrence
• What was the sequence of events that led up to the failure?
List changes, if any:
• Products
• Practices
• Procedures
• Environment
Who was involved?
• Personnel directly and indirectly involved
• Supervision and procurement/purchasing
What was the impact?
• Was anybody injured?
• Financial impact in terms of downtime, parts replacement, labor costs, and direct and indirect cost pools

Failure Mode (FM) Work Sheet

Note Physical Change of Part
Deformation
Fracture
Wear
Molecular transition (rust, corrosion, oxidation)
Other
Note Environmental Influence
Force type
Temperature
Time/age
Chemical
Other
Note Cadence
Steady
Random
Cyclic

Failure Mode (FM) Work Sheet (Continued)

Other
Note Articulation
• Amplitude (strong/weak)
• Frequency (fast/slow)
• Other
What part was affected?
Surface and work inward
Inside and work outward
Entire part, material, device, or tool all at once
Other

Failure Mode Work Sheet Key

Static (excessive elastic deformation, yield, buckling, collapse, etc.)
  Bulk material properties (yield strength, Modulus of Elasticity)
  Usually happens on first load application

Dynamic time dependent
  Corrosion
  Wear
  Stress corrosion cracking
  Chemical or product damage
  Creep
  Fatigue
  Weak link mechanism (importance of defects, statistical behavior)

Damage builds with time (some of the most catastrophic and costly in-service failures)
Sensitive to load levels, temperature, exposure time, number and rate of load cycles, environment (difficulties in testing and analysis)
Interaction effects (synergistic)

Process factors
- Loads
- Material
- Environmental factors
- Manufacturing and process variables

Conditions of local stress and strain

Chapter 4

Using the Information Gathered

In this chapter, information becomes the cornerstone of the specification. Data retrieved from various sources will be used to establish the requirements. The goal of Chapter 4 is to:

1. Identify the items with the greatest influence.
2. Identify the failure modes of items for proper performance selection.
3. Understand the physical properties and industry standards of typical items.

The *Law of Unintended Consequences* (author unknown) holds that almost all human actions have at least one unintended consequence. Unintended consequences, or situations where the final outcome is unexpected, can be classed into roughly three groups:

1. Positive, usually referred to as serendipity or a windfall
2. A source of problems, according to Murphy's law
3. Definitively negative or a perverse effect, which is the opposite result to the one intended

Up to this point, cost and value have been explained. Failure has been defined yet not completely investigated concerning specific items. To begin to write a comprehensive purchasing specification, one must begin to examine the documentation (including failure modes and properties) and begin assembling the foundation for the specification. One question that will be constantly asked is "What products should I write a specification for?" The specification is a list of requirements that address the particular concerns of your operation. Your operation is unique; therefore, it may require special attention. Gathering failure information is an important part of the specification development. It is actually an exercise that is relatively easy. It consists of listening to various technicians, mechanics, and even machine operators as they tell you what they think would work best and why. Before we get to the actual documentation process, we must first decide on what products actually require specifications.

Several hundred mechanics, technicians, and engineers were asked, "What items and materials do you buy on a routine basis that have a direct influence on productivity?" The following items had the greatest influence on productivity:

- Materials
  - Adhesives
  - Sealants
  - Paint and coatings
  - Gaskets, seals, and rings
  - Belts
  - Chain
  - Cable
  - Hose
  - Pipe
  - Lubricants
  - Filters
- Mechanical
  - Bearings
  - Gearing
  - Valves
  - Fittings
  - Bolts, screws, nuts
- Electrical
  - Wire
  - Lighting
- Motors

It seems fitting to look at these items as a basis for developing performance-based purchasing specifications. You are limited to these items. If you are looking to develop a purchasing specification of an item not listed above, take time to examine the process for the other items. This will provide a road map for your item of choice.

To develop a sound specification that drives reliability, it would be important to have a basic knowledge of that item as well as the ways in which it can fail. Once this information is established, then standards and specifications can be examined and written that would drive improvements that lead to profits. A great source for information would be your company's computerized maintenance management software (CMMS). Several years ago, I wrote a program that was Internet based and provided the user the ability to write work orders as well as close them out. The program also maintained a database of parts and the reasons why work was done. I named the program EMMA®, which was short for Equipment Maintenance Management Application. Truth be told, I named it after my daughter Emma. Hundreds of companies use this program, yet I have never actively marketed it. (It just seemed a little creepy when people would ask me to set them up with EMMA. A father has certain limitations!) Fortunately, many companies use a maintenance software product to keep track of all the repairs that take place as well as a log of all the replacement parts that have been used. By obtaining this information and then determining what the significant items are, the selection of what specifications to write first are based on a logical decision rather than on a whim.

Fact Finding and Documentation

One of the best ways to begin this process is to take a walk into the factory. Spend some time observing the operation. Depending upon the product produced, the environment can be very aggressive on parts and materials. It is a very good idea to understand the product's application. Below is an example of a total cost of failure (TCF) work sheet. It is a simple spreadsheet to build. As discussed in the "Summary" in Chapter 3, set up the sheet as follows:

	A	*B*	*C*	*D*	*E*	*F*
Incident and Reason	*Frequency*	*Item/ Part and Cost*	*Down-time Cost*	*Labor Costs*	*Additional Costs*	*Total Cost of Failure A * (B + C + D + E) = TCF*

If you were given the task of lowering costs, what item would you look to tackle first?

TCF Primary Work Sheet Example

	A	*B*	*C*	*D*	*E*	*F*
Incident and Reason	*Frequency*	*Item/ Part and Cost*	*Downtime Cost*	*Labor Costs*	*Additional Costs*	*Total Cost of Failure A * (B + C + D + E) = TCF*
Bearing failure on rollers due to premature bearings wear	96	$220 for bearing	4 hrs at $1000/hr = $4000 for each failure	4 labor hrs at $78/hr = $312 for each failure	Misshipment fines of $7045 annually	96 * ($220 + $4000 + $312) + $7045 = *$442,117*
Hydraulic pump failure from contam-ination—poor filters used	4	$3700 for pump, $25 for filter	8 hrs at $1000/hr = $8000 for each failure	16 labor hrs (2 man) at $78/hr = $1248 for each failure	50 gal of hydraulic oil = $900 ($18/gal) for each failure	4 * ($3725 + $8000 + $1248 + $900) = *$55,492*
Sensor misreads due to wire fry	129	$7 for wire	1 hr at $1000/hr	½ hr at $128/hr (electrician rate—billed for 1 hr)	$1,000,000 electrocution lawsuit settlement	129 * ($7 + $1000 + $128) + $1,000,000 = *$1,146,415*
Blower fan motor burnout	52	$387 for motor	2 hrs at $1000/hr = $2000	2 hrs at $128/hr (electrician rate) = $256	$89,937 in product spoilage	52 * ($387 + $2000 + $256) + $89,937 = *$227,373*

- The failure that costs the most would be the wire issue. At over one million dollars and a safety concern as well, you may think that is a project that should have your entire concentration.
- Consider the bearing failures. That may be a bit more complicated. Further investigation is in order. In the previous section, bearing failure was covered, and you learned that there are many reasons why those bearings fail. To make matters worse, they may not always be for the same reasons. Close to a half million dollars in costs sure makes the bearings an attractive target to go after.
- The blower fan motors would seem to be an easy fix as well. At least once a week a motor fails. At a total cost of failure of $227,373, the increased cost of a more robust motor would seem very logical provided that robustness can be defined and quantified. That leaves the hydraulic pump failures.
- The reason why the pumps have failed has been traced back to inferior filters being used. A $25 filter costs you over $50,000 annually. A finer filter may be

five times as expensive. A finer filter would trap the contaminants, resulting in longer pump life and reducing the total cost of failure. Would you spend $125 to save over $50,000?

So, which to repair or replace first? A bearing? A filter? A motor? A wire? All these maintenance, repair, and operations (MRO) items that many consider to be plain old off-the-shelf components, are readily available through any industrial distribution business. The problem is that the wrong items have always been bought. The failures have cost the company almost two million dollars.

I posed this question to several friends of various backgrounds and professions. Their responses are well thought out:

> At first glance it look as if the sensor misreads are *bleeding* the company dry based on your TCF analysis. However after closer inspection it looks as if the majority of that is based largely on assumption.
>
> After digging a little deeper on this, I would first attack to *premature bearing failure* to get a handle on cost overruns. This would be the simplest to remedy as I could do so by using a lubricant specifically formulated to address *failure.*
>
> By doing this, it should be relatively easy to extend the service intervals on the bearings, save downtime, free up maintenance to address the other issues, and at the end of the day results in savings of approx. $500,000.
>
> **Jerry Finn**
> *Vice President Sales, Mantek*

> The wires. Here is my reasoning: If the motors are getting the wrong commands because the information being used to command the motors is faulty, then the motors can be overstressed for the immediate application. Upgrading the wiring is a permanent fix while bearings are expected to eventually die. The hydraulic filters are nice but a serious retrofit will be necessary to install lower micron filters. The less the microns, the larger the filter housing to maintain the correct and necessary flow. Maintain the current micron size and get a filter cart for fine filtration.
>
> **Neil Novak**
> *Engineer and Former Maintenance Manager*

> My quick answer on the TCF example is to go back to my Danaher training and use DBS (Danaher Business System) principles. The core principles of DBS are *safety*, *quality*, *delivery*, and *cost*—in that order! Therefore, I would have to say the frayed wire issue would have to be Number 1 by far. Plus, from a moral standpoint, it just makes sense that you wouldn't want someone to get killed because you were too lazy or cheap to fix a known problem. With *quality* being Number 2, then the

blower issue has to come next. If product is being affected, you have to fix it. Just look at that Peanut Corp. problem. Your customers are your top priority, not your shareholders. Corporations have gotten it completely backwards in the last couple of decades, and that is why we are in the economic mess that we see today. If you take care of your customers first and your employees second, the shareholders will naturally reap the benefits of a sound company. Next is *delivery*—Here the bearing failures are the culprit. 384 hours per year of downtime due to bearing failures is ridiculous. You have to take care of this one or your customers are going to find someone else to make their sticky buns and ho-hos. Finally, we have *cost*—The filter example is a mistake that companies (and people in general) make all the time. Let's get the cheapest thing we can find. So what if we have to replace it five times as often as the alternative. If it is one-tenth the price, then it is a no-brainer, right? Well people do not see the true cost associated with going cheap until they crunch the numbers. Lost opportunity, lost productivity, intangibles, they all add up in the total cost of failure. But those costs are fuzzier and they do not translate well into balance sheets. When companies are run based on quarterly statements, it makes it hard to do the right thing for the longer-term.

Chris Aldrich
Owner, Mechanical Drive Components Inc.

I am really trying to get my brain wrapped around your question. However, not having all the information, I am having a really hard time fully understanding the problem. For example, how may sensors would need to be re-wired to insure no more misreads? How many hydraulic pumps are there that would need the more expensive filter? How many blower fan motors would need to be replaced with a more robust motor, and how much would the better motor cost? Is there a certain budget we have to work with? I see the number of failures for each item, but I am not sure if each of those failures is the same exact piece of equipment, or several different pieces of the same type of equipment.

I do not know if you can give us those answers, but based on the information given, my first inclination would be to repair the wiring issue depending on actually how many sensors would need to be rewired. The actual cost of repair (not total cost of failure) seems to be quite low with a tremendous return on investment (ROI).

Tim Meek
Owner and President DJL Postremo

On your TCF chart, some framing factors include: (1) *Goodwill cost of failure*—Goodwill is a business asset that needs to be included in any major contingency analysis. People may buy products that fail but are cheap, where the consequences of failure are minor and where replacement products/services are readily available. In some businesses, like hard-rock tunneling or mining, even loss of human life is an actuarial factor in the business model. But high financial or health risk is not tolerated in most businesses. So a single loss of life might put the business at risk; several might force a sale or bankruptcy. (2) *Regulatory aspects*—Beyond the financial aspects, failures with health and safety consequences can have major regulatory implications. This leads to a higher cost of doing business, possible restrictions on types of products or services, heavier oversight (with adverse effects on secrecy, for example) and even sanctions—all beyond (No. 1) above.(3) *Insurance and acceptance*—Sometimes insurance or self-insurance is part of the cost of doing business. Airlines have precise tables for the value of a human life, as do other industries that are exposed to rare but serious contingencies. Especially where loss of life or serious health problems is not likely, sometimes major screw-ups are accepted as cheaper than remediation.(4) *Normalization*—Back to the chart as it stands, I would find it hard to compare priorities without a cost normalized to time (years), sales (cost per $ million), staff (cost per operator) or some other business-relevant metric. A critical factor would be the recursive effect of any decision: for example, remediation will reduce the cost or frequency of each accident, presumably more than the cost of remediation. So the statistics will change. Even the "accept occasional failures" choice leads to greater efficiencies in damage control. This is nominally a second-order effect, but might dominate the whole decision process.

Alaric Naiman
PhD, Chemistry

My first question would be how does each of these fit within the company's mission statement? They are all mechanical problems, so they may all be within the scope of the company's mission statement.However, with that said, there appears to me, to be one fix that is going to be required no matter what. The wire problem has safety issues and has already developed into a lawsuit. Not acting on the knowledge that you are putting people's safety at risk, could be a lot more costly than the current $1,000,000 lawsuit. Management knows of the problem and knows that there is a safety issue. Should another accident occur due to this problem, inaction not only jeopardizes the health of employees, but also magnifies the chance of an even greater lawsuit. Now, this action is reactive rather than proactive and the company needs to become more pro-active in their approach. But in this situation, the company needs to address the one issue that jeopardizes the safety of its employees. Also, the precise reason for the bearing failures need to be determined before a decision is made for further action to any of the other issues. Also, if not already in place, there should be a quality review team that conducts ongoing reviews of all mechanical components. This effort could help catch problems early on and minimize the ultimate costs of repairing or replacing parts. I am sure there are other approaches that can be justified as well. But with the information provided, to me, this appears the best approach to take.

Glen Scriber
Client Relocation, Prudential Insurance

From the information, I cannot compare apples to apples, not knowing how much labor is involved as well as detailed information on each problem. From what I can see, I would pick filter replacement. The man power would be the same, no intense labor to factor in, and the ratio from maintenance cost to savings is great.

I do not know cost to savings ratios on all of the other items, just what it cost them each year, not a cost for options comparison. I would understand it much better if there were a (1) problem and (2) total cost to fix the problem and (3) savings estimate for fixing the problem (how much money saved each year) for each and every problem presented below per year. It is kind of like a word problem and you have to have all of the components before you are able to solve the problem. I do not see all three components for each issue below.

Paula Coco Cuva
Photography Stylist

One observation that you may have made is that safety is of high concern. It is one thing to lose customers due to failure; it is another matter if one of your employees or contractors gets hurt or worse. All companies have their own vision statement and operating procedures. The company that thrives is one that cares for its employees.

I intentionally made the example vague. Often in industry, record keeping can be rather lacking. Details of incidents are often the stuff of good crime fiction with an emphasis on fiction. This is normally due to the highly suspicious nature of employees. Although few will acknowledge their paranoia, the majority will take great pains at figuring out the angle that leaves them least vulnerable for disciplinary action due to incompetence or hasty behavior. It is truly essential that the fact-finding missions be prefaced by a guarantee of anonymity. It may be difficult to understand the level of paranoia that exists, but rest assured that it is a strong force in industry.

Physical Properties and Standards

To write a comprehensive purchasing specification that will help drive reliability, it is often necessary to have materials and devices tested in order to understand the performance and physical properties under different physical demands. Only in rare cases will a buyer actually have an item tested for a particular characteristic. Normally, the manufacturer will have the item tested according to what the manufacturer feels is most required. Often the manufacturer will run the minimum amount of tests or will not publish the results for fear of tipping the hand of a competitive edge or out of laziness. In any case, it is best to provide a specification for the supplier to respond to.

A performance-based procurement specification normally takes several weeks to prepare. After the maintenance and engineering staff performed an failure mode effect analysis (FMEA), they contacted several vendors to document the performance criteria for each product as it related to the various failure modes (which were identified from the failed parts). Many times, suppliers use ASTM (American Society for Testing and Materials) standards or other industry standards. ASTM has standards for just about any material or product. These are typically industry standards. These standards provide the foundation for performance levels of many products.

ASTM International is a global forum for the development of consensus standards. Organized in 1898, ASTM is one of the largest voluntary standards developing organizations in the world. ASTM is a not-for-profit organization that provides a forum for the development and publication of voluntary consensus standards for materials, products, systems, and services. ASTM's members, representing producers, users, consumers, government, and academia from over 100 countries, develop technical documents that are a basis for manufacturing, management, procurement, codes, and regulations. These members belong to one or more committees, each of which covers a subject area such as steel, petroleum, medical devices, property management, consumer products, and many more. These committees develop the more than 11,000 ASTM standards that can be found in the 77-volume *Annual Book of ASTM Standards*. For over a century, industries around the world have turned to ASTM International for the development of voluntary consensus standards. Known for their high technical quality and market relevance, these standards provide an important part of the information infrastructure that guides design, manufacturing, and trade in the global economy. An online index of 11,000 ASTM standards enables you to locate ASTM standards in 130 varying industry areas. Available on the ASTM Web site (www.astm.org), the online index facilitates searches by keyword or standard number, and viewers can access the titles and scopes of all ASTM standards. The full text of any ASTM standard is available electronically or in print via the Web site or through customer service at ASTM International (Phone: 610-832-9585). Your vendors should be well versed in their particular standards for their products.

A list of the common standards for every conceivable item is also available through the American National Standards Institute (ANSI). The ANSI accredits qualified organizations to develop standards in the technical area in which they have expertise. ANSI's role is to administer the voluntary consensus standards system, providing a neutral forum for the development of policies on standards issues and serving as an oversight body to standards development and conformity assessment programs and processes.

The ANSI is a private, nonprofit organization whose mission is to enhance U.S. global competitiveness and the American quality of life by promoting, facilitating, and safeguarding the integrity of the voluntary standardization and conformity assessment system.

There are several organizations that aid in the development and distribution of standards used in industry. The International Organization for Standardization (ISO) has more than 800 technical committees and subcommittees involved in standards development. The United States participates in more than 600 of these committees. This list contains all of the ISO technical committees and subcommittees. If the United States is active in an activity, the name of the U.S. standards developing organization (SDO) that serves as the Technical Advisory Group (TAG) administrator is provided in the detailed listing.

The International Electrotechnical Commission (IEC) has more than 169 technical committees and subcommittees involved in standards development. The United States participates on more than 152 of these committees. This list contains all of the IEC technical committees and subcommittees. If the United States is active, the name of the national committee member will be listed in the International Electrotechnical Commission (USNC) and the approved Technical Advisory Group (TAG).

The U.S. Standards Strategy (USSS) helps to advance trade issues in the global marketplace, enhance consumer health and safety, and meet stakeholder needs.

Obtaining Standards from the American National Standards Institute (ANSI)

Table 4.1 is a comparison table summarizing at a glance the various ways that you can obtain the standards you need.

Tables 4.2 and 4.3 list publishers of various standards.

Table 4.1 Obtaining ANSI Standards

	Individual Standards	*Packages*	*eSubscriptions*	*Site License*	*NSSN*
One or More User Licenses	x	x	x	x	x
Volume Discounts		x	x	x	x
Multiple Downloads of Same Standard (See Rights)			x	x	x
Updates			x	x	x

Note: Updates are new standards provided to replace or supplement any documents in the collection that change.

Table 4.2 ANSI Standards Details

Product	*Need*	*Rights**	*Ordering*	*Availability*
Individual Standards	One or just a few licenses for each standard, one-time download	One download during 7-day term; right to use one electronic copy and one print copy for one individual per license	Online: webstore.ansi.org	All standards
Packages	One or just a few licenses for each standard, one-time download, related standards in a convenient group offered at a discount	One download during 7-day term; right to use one electronic copy and one print copy for one individual per license	Online: webstore.ansi.org	Check online
eSub	Up to 20 simultaneous user licenses, volume discount, unlimited download of standards plus download updates during subscription term; e-mail alerts for new and revised standards; e-mail alert for renewal at end of term	Unlimited downloads during a 1-year period; right to use one electronic and one paper copy for each unique user; discount price on predefined collections; limit of 50 standards or less for custom, cherry-picked collections; no additional charge for updates	Online: webstore.ansi.org/subscriptions	Check online

(*Continued*)

Table 4.2 ANSI Standards Details (Continued)

Product	*Need*	*Rights**	*Ordering*	*Availability*
Site License	Unlimited download, volume discount, frequent use, several concurrent users	Unlimited access during term of license; right to use electronic and paper copies for each licensed site; substantial discount versus individual purchase; automatic update notification; documents hosted on customer server	Contact customer support	Call for availability
NSSN	Unlimited downloads, volume discount, full collections, frequent use, many concurrent users, managed network infrastructure	Search functionality and fee-based optional custom integration; unlimited access during term of license; right to use unlimited electronic and paper copies for each licensed site; substantial discount versus individual purchase; updates added automatically	Contact customer support	Call for availability

Table 4.3 Publishers of Various Standards

AAMI	Association for the Advancement of Medical Instrumentation
ABMA	American Bearing Manufacturers Association
ABNT	Brazilian Technical Standards Association
ABYC	American Boat and Yacht Council
ACC	American Chemistry Council
ACHC	Accreditation Commission for Health Care
ADA	American Dental Association
AGA	American Gas Association
AGMA	American Gear Manufacturers Association
AHAM	Association of Home Appliance Manufacturers
AIAA	American Institute of Aeronautics and Astronautics
AIAG	Automotive Industry Action Group
AIHA	American Industrial Hygiene Association
AIIM	Association for Information and Image Management
AMT	The Association for Manufacturing Technology
APSP	Association of Pool and Spa Professionals
APTA	American Public Transportation Association
ARI	Air-Conditioning and Refrigeration Institute
ARMA	Association for Information Management Professionals
ASA	Acoustical Society of America
ASABE	American Society of Agricultural and Biological Engineers
ASHRAE	American Society of Heating, Refrigerating and Air-Conditioning Engineers
ASME	American Society of Mechanical Engineers
ASQ	American Society for Quality
ASSE	The American Society of Safety Engineers
ASTM	American Society for Testing and Materials International

(*Continued*)

Table 4.3 Publishers of Various Standards (Continued)

ATIS	Alliance for Telecommunications Industry Solutions
AWS	American Welding Society
BHMA	Builders Hardware Manufacturers Association
CEMA	Conveyor Equipment Manufacturers Association
CGA	Compressed Gas Association, Inc.
CITRA	Center for International Regulatory Assistance
CLSI	Clinical and Laboratory Standards Institute
CSA	Canadian Standards Association
DIN	DIN Deutsches Institut für Normung
DOD	U.S. Department of Defense
ESTA	Entertainment Services and Technology Association
ETSI	European Telecommunications Standards Institute
GEIA	Government Electronics and Information Technology Association
GISC	Glazing Industry Secretariat Committee
GTEEMC	Georgia Tech Energy and Environmental Management Center
HL7	Health Level Seven
IACET	International Association for Continuing Education and Training
IAPMO	International Association of Plumbing and Mechanical Officials
ICBO	International Conference of Building Officials
ICC	International Code Council
IEC	International Electrotechnical Commission
IEEE	Institute of Electrical and Electronics Engineers
IESNA	Illuminating Engineering Society of North America
IESO	Indoor Environmental Standards Organization
IFI	Industrial Fasteners Institute
IICRC	Institute of Inspection, Cleaning and Restoration Certification
I3A	International Imaging Industry Association

Table 4.3 Publishers of Various Standards (Continued)

INCITS	International Committee for Information Technology
IPC	Association Connecting Electronics Industries
ISA	Instrument Society of America
ISO	International Organization for Standardization
JIS	Japanese Industrial Standards
KOK	Metric eBooks
LIA	Laser Institute of America
MSS	Manufacturers Standardization Society of the Valve and Fittings Industry, Inc.
MTS	Institute for Market Transformation to Sustainability
NACE	National Association of Corrosion Engineers
NADCA	North American Die Casting Association
NAESB	North American Energy Standards Board
NASPO	North American Security Products Organization
NECA	National Electrical Contractors Association
NEMA	National Electrical Manufacturers Association
NETA	InterNational Electrical Testing Association
NFPA	National Fire Protection Association
NFPA	National Fluid Power Association
NISO	National Information Standards Organization
NPES	National Printing Equipment and Supply Association, Inc.
NPPC	National Pork Producers Council
NSF	NSF International
OEOSC	Committee for Optics and Electro-Optical Instruments
OLA	Optical Laboratories Association
ON	Austrian Standards Institute
OPEI	Outdoor Power Equipment Institute

(Continued)

Table 4.3 Publishers of Various Standards (Continued)

PCC	ANSI Partially Controlled Collections
PMI	Project Management Institute
PSDA	Print Services and Distribution Association
QuEST	QuEST Forum: Quality Excellence for Suppliers of Telecommunications
SAI	Standards Australia
SCTE	The Society of Cable Telecommunications Engineers
SEMI	Semiconductor Equipment and Materials International
SES	Standards Engineering Society
SIA	Scaffold Industry Association
SIA	Security Industry Association
SIS	Swedish Standards Institute
SPC	CHINESE Standards
TAPPI	Technical Association for the Pulp, Paper, and Converting Industry
TCA	Tile Council of North America
WMMA	Wood Machinery Manufacturers Association

Failure analysis has been able to help identify the physical properties that will help build your specification. Performance tables for various MRO materials are offered here as a guide and not as the final say in the performance section of the specification.

The following is a summation of the various failure modes for products that generally have a direct influence on reliability. The failure modes have been retrieved from interviews as well as documentation. By understanding why items fail, we can look to purchase items that are designed to last longer and hold up better in the various conditions typical to your facility. If your facility has equipment that constantly is washed down with a caustic solution and you consistently have to purchase equipment coatings because the washdown chemical is eating away at the coating, perhaps savings could be appreciated if a caustic-resistant coating was purchased instead. The only way to know is to set a specification and then examine what the offerings are that best suit your requirements of caustic resistance.

Included in each section is a table of physical properties that addresses performance and failure relief. Also included are tables of various test standards that consider a generalized approach for suppliers to test their offerings. These tables can be used as a foundation for the performance requirements of the specification.

Adhesives and Sealants

Failure is the result of two separate phenomena: adhesive failure (when the material releases from the substrate) and cohesive failure (when the material breaks apart internally from itself). There are a number of reasons why these actions occur.

Several years ago, a study was conducted on product complaints received for residential adhesives, and only a few occurred with any regularity:

- Adhesive does not stick as expected, usually in vertical applications.
- There is a lack of adhesion, if any.
- The adhesive or sealant never cures.
- There are bubbles that show up on the surfaces of the materials being bonded.

It would stand to reason that the typical failures experienced are the result of the following failure modes: substrates, environment, and adhesive chemistry. The elements of proper adhesion with minimal chance for adhesive failure include:

- Clean, solid substrate
- Substantially dry surface
- Reasonably flat surface
- Matching adhesive chemistry with the application

The two predominant mechanisms of failure in adhesively bonded joints are adhesive failure or cohesive failure. Adhesive failure is the interfacial failure between the adhesive and one of the surfaces. It indicates a weak-boundary layer often from improper surface preparation or adhesive choice. Cohesive failure is the internal failure of either the adhesive or, rarely, one of the substrates. Ideally, the bond will fail within one of the substrates. This indicates that the maximum strength of the bonded materials is less than the strength of the adhesive strength between them. Usually, the failure of joints is neither completely cohesive nor completely adhesive. Measurement of the success of a particular joint is based on the relative percentage of cohesive failure to adhesive failure. The precise cause of premature failures is difficult to determine. The variety of causes include adverse stresses (peeling, for example), rate of application of stresses, fatigue, temperature, humidity, and solvents. Many of the failure modes for adhesive and sealant failure are the same for paints and coatings as well as other materials.

Adhesive Physical Properties

Material system
Chemical system
Filler
Cure/technology
Features and industry/applications
Type/form
Features
Substrate compatibility
Industry
Material properties (nominal/typical)
Process and physical
Viscosity (centipoises [cP])
Gap fill (inch)
Thermal
Use temperature (°F)
Thermal conductivity (watt per meter Kelvin [W/m-K])
CTE (microinch [μin.]/in-F)
Mechanical
Tensile (break) (pound/square inch [psi])
Elongation (%)
Electrical
Resistivity (ohm-cm)
Dielectric strength (kilovolt [kV]/in.)
Dielectric constant (decibal [dB]/kilometer [km])
Optical
Index of refraction
Transmission (%)

Sealant Physical Properties

Material system
Chemical system
Filler
Cure/technology
Type/form
Substrate compatibility
Features and industry/applications
Features
Industry
Material properties (nominal/typical)
Process and physical
Viscosity (cP)
Gap fill (inch)
Thermal
Use temperature (°F)
Thermal conductivity (W/m-K)
CTE (μin/in-F)
Mechanical
Tensile (break) (psi)
Elongation (%)
Electrical
Resistivity (ohm-cm)
Dielectric strength (kV/in)
Dielectric constant
Optical
Index of refraction
Transmission (%)

Adhesive and Sealant Properties and Test Methods/Standards

Property	*ASTM Test Methods and Standards*
Aging	D1183 Practices for Resistance of Adhesives to Cyclic Laboratory Aging Conditions
	D3632 Test Method for Accelerated Aging of Adhesive Joints by the Oxygen-Pressure Method
Chemical Resistance	D896 Practice for Resistance of Adhesive Bonds to Chemical Reagents
Cleavage Strength	D1062 Test Method for Cleavage Strength of Metal-to-Metal Adhesive Bonds
Corrosive Resistance	D3310 Test Method for Determining Corrosivity of Adhesive Materials
Creep Resistance	D2293 Test Method for Creep Properties of Adhesives in Shear by Compression Loading (Metal-to-Metal)
	D2294 Test Method for Creep Properties of Adhesives in Shear by Tension Loading (Metal-to-Metal)
Environmental Resistance	D1151 Practice for Effect of Moisture and Temperature on Adhesive Bonds
Flash Point	D1310 Test Method for Flash Point and Fire Point of Liquids by Tag Open-Cup Apparatus
	D3278 Test Methods for Flash Point of Liquids by Small Scale Closed-Cup Apparatus
	D56 Test Method for Flash Point by Tag Closed Tester
Flexibility	D1184 Test Method for Flexural Strength of Adhesive Bonded Laminated Assemblies
	D3111 Test Method for Flexibility Determination of Hot-Melt Adhesives by Mandrel Bend Test Method
Flow	D1084 Test Methods for Viscosity of Adhesives
Fracture Strength	D5041 Test Method for Fracture Strength in Cleavage of Adhesives in Bonded Joints
Gap Fill	D3931 Test Method for Determining Strength of Gap-Filling Adhesive Bonds in Shear by Compression Loading
Hardness	D2240 Test Method for Rubber Property—Durometer Hardness

Adhesive and Sealant Properties and Test Methods/Standards (Continued)

Property	*ASTM Test Methods and Standards*
Impact Strength	D950 Test Method for Impact Strength of Adhesive Bonds
Low-Temperature Service	D2557 Test Method for Tensile-Shear Strength of Adhesives in the Subzero Temperature Range from –267.8 to –55°C (–450 to –67°F)
Peel Strength	D1876 Test Method for Peel Resistance of Adhesives (T-Peel Test)
	D2918 Practice for Durability Assessment of Adhesive Joints Stressed in Peel
	D3167 Test Method for Floating Roller Peel Resistance of Adhesives
	D3807 Test Method for Strength Properties of Adhesives in Cleavage Peel by Tension Loading (Engineering Plastics-to-Engineering Plastics)
Pumpability	D5267 Test Method for the Determination of Extrudability of Cartridge Adhesives
Shear Strength	D1002 Test Method for Apparent Shear Strength of Single-Lap-Joint Adhesively Bonded Metal Specimens by Tension Loading (Metal-to-Metal)
	D3166 Test Method for Fatigue Properties of Adhesives in Shear by Tension Loading (Metal/Metal)
	D3528 Test Method for Strength Properties of Double Lap Shear Adhesive Joints by Tension Loading
	D5656 Test Method for Thick Adherent Metal Lap-Shear Joints for Determination of the Stress–Strain Behavior of Adhesives in Shear by Tension Loading
	D5868 Test Method for Lap Shear Adhesion for Fiber Reinforced Plastic (FRP) Bonding
	D905 Test Method for Strength Properties of Adhesive Bonds in Shear by Compression Loading
	E229 Test Method for Shear Strength and Shear Modulus of Structural Adhesives

(*Continued*)

Adhesive and Sealant Properties and Test Methods/Standards (Continued)

Property	*ASTM Test Methods and Standards*
Storage Life	D1337 Practice for Storage Life of Adhesives by Consistency and Bond Strength
Stress Cracking	D3929 Test Method for Evaluating Stress Cracking of Plastics by Adhesives Using the Bent-Beam Method
Temperature Stability (Hot Melt)	D4498 Test Method for Heat-Fail Temperature in Shear of Hot Melt Adhesives
	D4499 Test Method for Heat Stability of Hot-Melt Adhesives
Tensile Strength	D3983 Test Method for Measuring Strength and Shear Modulus of Nonrigid Adhesives by the Thick-Adherent Tensile-Lap Specimen
	D897 Test Method for Tensile Properties of Adhesive Bonds
Torque Strength	D5648 Test Method for Torque-Tension Relationship of Adhesives Used on Threaded Fasteners (Lubricity)
	D5649 Test Method for Torque Strength of Adhesives Used on Threaded Fasteners

Paints and Coatings

- Surface condition
- Surface preparation
- Humidity/dew point
- Coating chemistry
- Applicator skill
- Curing conditions
- Transport and handling

Alligatoring: The patterned cracking in the surface of the paint film resembling the regular scales of an alligator generally found on wood surfaces. Alligatoring occurs when paint cannot adhere to a glossy surface, when a second coat of paint is applied over an inadequately dried first coat of paint, weather aging, excessive coats of paint, or when the finish coat expands and contracts at a greater extent than any underlying coats. Paint must be completely removed when alligatoring has occurred to ensure an even and uniform finish. Wood surfaces should be primed with a top-quality alkyd primer and acrylic latex finish. These are used on an application of an extremely hard, rigid coating, like alkyd enamel, over a more flexible coating, such as a latex primer.

Alligatoring can also occur after application of a topcoat before the undercoat is dry. It can also be caused by the natural aging of oil-based paints as temperatures fluctuate. The constant expansion and contraction result in a loss of paint film elasticity.

Bleeding: An unsightly surface discoloration commonly found on exterior wood surfaces and hardboard siding, ceilings, repainted wallpaper, or when light-colored paint is applied over dark-colored paint. The causes of bleeding are moisture and water-soluble dyes located within wood surfaces and wallpaper, and inadequate priming of surfaces. Bleeding on hardboard siding is caused by wax. The removal of any excessive water source is essential before any surface preparation can be completed. Exterior wood surfaces should be primed with top-quality alkyd-based primers and finished with top-quality acrylic latex paints. Interior surfaces should be primed with either a top-quality alkyd-based or acrylic latex–based stain blocking primer, and finished with top-quality acrylic latex finish paint.

Blistering: Occurs with fresh paint applied in direct sunlight, and moisture causes paint to blister. Surfaces that are heated by the sun and become hot can prematurely dry new paint, causing blistering. This is more common with dark or dramatic colors, because they, in addition to surfaces, absorb heat. Heat blistering will have a layer of paint under the blister if the surface is being repainted. Trapped moisture in walls or behind surfaces will eventually try to escape through painted surfaces, also causing blistering. Moisture blistering will reveal a bare surface under the blister. Blistering surfaces should be scraped and sanded smooth, primed with a top-quality alkyd or acrylic latex primer, and finished with top-quality acrylic latex finish paint. If moisture blistering occurs on a repainted surface, spot priming of bare spots will be necessary to ensure a uniform and even finish before the final priming and painting are completed. Blistering results from localized loss of adhesion and lifting of the paint film from the underlying surface. This occurs from painting in direct sunlight or on a surface that is too warm, especially when applying a dark-colored solvent-based coating; applying an oil-based or alkyd paint over a damp or wet surface; excess humidity or other moisture escaping from inside through the exterior walls (less likely with latex paints, which allow water vapor to escape without affecting the paint film); and exposing a latex paint film to excessive moisture in the form of dew, high humidity, or rain shortly after the paint has dried, especially if there was inadequate surface preparation or a lower-quality paint was applied.

Burnishing: The increase in gloss or sheen of paint film when subjected to rubbing, scrubbing, or having an object brush up against it. This failure may occur due to the use of flat paint in highly trafficked areas, where a higher sheen level would be desirable, and with frequent washing and spot cleaning. This can also occur due to the use of lower grades of paint with poor stain and scrub resistance.

Chalking: The weathered powder found on exterior painted surfaces; this is a normal way for paint to weather away. However, premature chalking can be caused by inadequate priming or thin paint. Technicians should be aware that chalking on siding located above masonry will run during rainy weather, causing the masonry to stain. Once a chalked surface is pressure-washed clean and prepared for painting, one coat of a top-quality alkyd primer and one coat of a top-quality acrylic latex finish paint are highly recommended to prevent future excess chalking.

Checking: An early form of cracking; this occurs when paint begins to lose elasticity, forming narrow breaks on the painted surface. These breaks later swell and become cracks. Checking and cracking are found on wood surfaces with multiple layers of paint, or plywood. Checked or cracked paint should be completely removed from surfaces before repainting. Properly prepared wood surfaces should be primed with a top-quality alkyd primer and finished with top-quality acrylic latex finish paint.

Cracking/Flaking: The splitting of a dry paint film through at least one coat as a result of aging, which ultimately will lead to complete failure of the paint. In its early stages, the problem appears as hairline cracks; in its later stages, flaking occurs. This occurs because of lower-quality paint that has inadequate adhesion and flexibility and is due to overthinning or overspreading the paint. This can also occur due to inadequate surface preparation or applying the paint to bare wood without first applying a primer and to excessive hardening and embitterment of alkyd paint as the paint job ages.

Efflorescence: A white, salty substance that forms on masonry and plaster due to moisture migrating through the surface. If left untreated for long periods of time, it will become hard and crusty. Any source of excessive moisture must be eliminated, the surface must be cleaned by scraping or wire brushing, and surfaces must be repaired where applicable before priming and painting. Surfaces can be painted with top-quality alkyd or acrylic latex–based primers and paints that are alkaline resistant.

Fading: Caused by natural weathering due to the ultraviolet rays of the sun. Chalking pigment, alkali from masonry, and tint intended for interior-based paints but used on exterior surfaces will cause paint to fade prematurely.

Foaming/Cratering: From the formation of bubbles (foaming) and resulting small, round concave depressions (cratering) when bubbles break in a paint film, during paint application and drying. This typically occurs because of shaking a partially filled can of paint, low-quality paint or very old latex paint, applying (especially rolling) paint too rapidly, or the use of a roller cover with wrong nap length. It also occurs due to excessive rolling or brushing of the paint or applying a gloss or semigloss paint over a porous surface.

Hatbanding: Caused by the over cutting in of interior walls, ceilings, and trim corners, and the use of excessively long roller naps. It occurs when wet paint is applied over dry paint. Hatbanding is very noticeable with dark and dramatic interior colors and sheens. Using small nap rollers and applying a thin brush coat of paint will help eliminate hatbanding.

Lap Marks: Due to the inadequate application of coatings. They are mostly seen on exterior surfaces. If a technician or contractor is brush painting siding, the paint must be applied horizontally from one end to the other without stopping. This also occurs when the application of the paint to the surface is continued; a lap mark forms where the paint overlapped. This is a major reason why coatings should never be applied in direct sunlight. Coatings should be applied in shady areas in warm months, and technicians and contractors should stay ahead of the sun in cooler months. Staying ahead of the sun will allow it to come around and dry the area most recently painted.

Mildew: A fungus that grows best in a warm, humid climate and is often found in shaded areas but is not limited to shaded areas. Mildew can appear in a variety of colors and prefers painted surfaces because of the nutrients that paints provide. These same nutrients float through the air and land on painted surfaces and become food. Mildew grows faster on flat surfaces than on glossy surfaces and is attracted to linseed oil–based products that do not contain a biocide. Most paints and stains come already made with a biocide included. However, a biocide can be added to paints and stains that do not contain one. Mildew can be effectively cleaned with a 3-to-1 ratio of water to bleach. Proper surface preparation and cleaning are crucial for mildew prevention, in addition to using top-quality primer and acrylic latex finish paint. Acrylic latex finish paints are the most mildew resistant. Painting contractors who perform quality surface preparation and painting are a technician's best ally in preventing the growth of mildew.

Paint Incompatibility: Results in the loss of adhesion where many old coats of alkyd or oil-based paint receive a latex topcoat. Use of water-based latex paint over more than three or four coats of old alkyd or oil-based paint may cause the old paint to "lift off" the substrate.

Peeling: Occurs on a variety of surfaces and is directly linked to moisture and inadequate surface preparation. It is often seen as the spontaneous loss of ribbons or sheets of paint due to loss of adhesion. When the surface is coated with a primer and topcoat, or with several coats of paint, the peeling may involve all the coats, some of them, or just the topcoat. It can occur due to swelling of wood due to seepage or penetration of rain, humidity, and other forms of moisture into the home through uncaulked joints, deteriorated caulk, leaking roof, or other areas of excess humidity or other moisture escaping from within the home through the exterior walls (less likely with latex paints, which allow water vapor to escape without affecting the paint film); inadequate surface preparation; use of a lower-quality paint that has inadequate adhesion and flexibility characteristics; applying latex paint under conditions that hinder good film formation (e.g., on a very hot or cold day; in windy weather applying an oil-based paint over a damp or wet surface).

Understanding different surfaces and how they react to moisture and coatings will greatly reduce the risk of this common paint problem.

Picture Framing: The effect of nonuniform color that can appear when a wall is painted with a roller but is brushed at the comers. The brushed areas generally appear darker, resembling the "frame" of a "picture." Also, sprayed areas may be darker than neighboring sections that are brushed or rolled. Picture framing can also refer to sheen effects. This is from a hiding (coverage) effect. Brushing will generally result in lower spread rates than rolling, producing a thicker film and more hiding. This also occurs when adding colorant to a nontintable paint or using the wrong type or level of colorant.

Poor Hiding: The failure of dried paint to obscure or "hide" the surface to which it is applied from the use of low-quality paint, low-quality tools/wrong roller cover, improper combination of tinting base and tinting color, poor flow and leveling. This is also due to use of a paint that is much lighter in color than the substrate or that primarily contains low-hiding organic pigments or application of paint at a higher spread rate than recommended.

Poor Stain Resistance: The failure of the paint to resist absorption of dirt and stains from the use of a lower-quality paint and from applying paint to an unprimed substrate.

Rusting: Occurs when metal is exposed to moisture and oxygen. Any metal surfaces should be completely sealed with a rust-inhibited primer. If rust appears through painted surfaces, it should be wire brushed to bare metal or treated with a coating that will harden it. Two coats of a top-quality rust-inhibited alkyd or acrylic latex primer should be applied to metal surfaces once the rust is treated or wire brushed away. Rusty nails that bleed through surfaces should be reset, coated with a top-quality rust-inhibited primer, caulked, primed, and painted with top-quality coatings.

Sagging and Running: Occur when paint is applied over glossy surfaces, there is excessive thinning, there is too much paint, or paint is being applied to a dirty surface. Proper application of paints and surface preparation will eliminate sags and runs.

Surfactant Leaching: Concentration of water-soluble ingredients on the surface of a latex paint, typically on a ceiling surface in rooms that have high humidity (e.g., shower, bathroom, kitchen); may be evident as tan or brown spots or areas, and can sometimes be glossy, soapy, or sticky. All latex paint formulas will exhibit this tendency to some extent if applied in areas that become humid (bathrooms, for example), especially in ceiling areas.

Wrinkling: A rough, crinkled paint surface that occurs when uncured paint forms a "skin." This can occur when paint is applied too thickly (more likely when using alkyd or oil-based paints). Hot weather or cool, damp weather, which causes the paint film to dry faster on top than on the bottom, can be a culprit. High humidity is often a source of failure unless the material is designed with that in mind, as is a topcoat of paint to insufficiently cured primer. Another source is painting over a contaminated surface such as dirt, dust, or wax.

Coating Failures

Coating failures may appear during application, at the stage of curing, or after a certain period of service life. Statistics show that as much as 95% of all coating failures are a result of poor surface preparation and application.

Below you will find some examples of common coating failures and the reasons why they occur. Please note that there may be many reasons for a coating failure, and, in some cases, it requires a lot of experience to find the exact cause.

Sagging (occurs when):

- Paint is applied in excess of the dry film thickness (DFT) specified.
- Too much thinner has been added to the paint.
- The gun is held too close to the surface.
- Sags are recognized as "curtains" on painted surfaces. If the wet film thickness is too high, excessive sagging can result in pools of paints forming on horizontal surfaces or in corners. After curing, the paint may crack all the way to the substrate in such areas and reveal unprotected steel. If sagging is noticed at the spraying stage, it should be brushed out while the paint film is still wet.

Pinholes and Pores: Occur when using the wrong spraying technique, such as excessive air pressure, excessive film thickness, strong wind (too good ventilation), and too long application distance, causing craters, pinholes, and pores. If noticeable on the paint film, check the spraying equipment to ensure that the air pressure and nozzle size are correct. Pinholes in a paint film can also result from overspray. On excessive film thickness, air will be trapped in the paint. The escaping air will create pinholes. The consequence is that pinpoint rusting occurs, followed by undercutting of the coating around the pinholes.

Blistering: This is one of the most common types of failure related to the adhesion of the paint. Sometimes the blisters are dry, and sometimes they are filled with liquid. The blisters can be both large and small, often shaped as hemispheres. The size usually depends on the degree of adhesion to the substrate, or between the coats, and the internal pressure of the gas or the liquid inside the blister. Blistering can be caused by a number of different conditions:

- Soluble salts contaminating the substrate or contaminating the surface between coats. No coatings are 100% waterproof. The moisture vapor passing through the coating can dissolve salt into a concentrated solution. Pressure in the high-concentration liquid will cause blisters. This phenomenon is called osmosis.

- Contamination of the surface (for example, oils, waxes, dust, etc.) will not allow proper adhesion of the coating. The moisture vapor tends to be concentrated in these areas of low adhesion. In this case, the blisters are "dry" blisters.
- Poor or inadequate solvent release from the coating. Entrapped solvents can increase the water absorption and moisture vapor transmission of the coating and lead to blistering. Solvent odor is usually connected with retained solvents. If the blistering is widespread on a construction, remove all flakes and wash before a new system is applied. For local areas, blast or carry out other mechanical cleaning before recoating.

Lifting: A rising of the undercoat. It is caused by a stronger solvent in the topcoat attacking the previously applied film. The result is a wrinkled surface. A typical example is a topcoat containing xylene, on top of an alkyd-based primer containing white spirit. The xylene in the topcoat will dissolve the primer.

Delamination/Peeling: The loss of adhesion to the substrate or between coats of paint is delamination or peeling. The causes include:

- Unsatisfactory surface preparation
- Incompatible primer or undercoat
- Substrate or intercoat contamination
- Excessive cure time between coats

Orange Peel: Finely pebbled or dimpled surface texture with an appearance similar to the skin of an orange. Caused by:

- Improper atomization due to low air pressure
- Spraying too close to the surface
- Rapid solvent evaporation

Paint and Coating Physical Properties

Chemistry
Cure/setting technology
Coverage and thickness
Coverage (ft^2)
Coating thickness
Wet film (mil)
Dry film (mil)
Processing
Viscosity (cps)

Paint and Coating Physical Properties (Continued)

Specific gravity
Cure temperature (°F)
Cure time (minutes)
Pot life/application time (minutes)
Humidity (%)
VOC (lbs/gal)
Form
Dispensing method
Operating temperature (°F)
Color
Finish
Conductivity (μS/cm)
Resistivity (ohm-cm)
Dielectric strength (volt per meter [V/m])
Functional features

Paint and Coating Standards and Test Methods

Property	*ASTM Test Methods and Standards*
Abrasion Resistance	D5181 Test Method for Abrasion Resistance of Printed Matter by the GA-CAT Comprehensive Abrasion Tester
Bleed Resistance	Bleeding—ASTM D 279-02 Standard Test Method for Bleeding of Pigments
	Blistering—ASTM D 714-02 Standard Test Method for Evaluating Degree of Blistering of Paints
Burnish Resistance	Burnishing—ASTM D 6736-08 Standard Test Method for Burnish Resistance
Checking Resistance	Checking—ASTM D 660-93 Standard Test Method for Evaluating Degree of Checking

(*Continued*)

Paint and Coating Standards and Test Methods (Continued)

Property	*ASTM Test Methods and Standards*
Crack Resistance	Cracking/Flaking—ASTM D 661-93 Standard Test Method for Evaluation of Degree of Cracking, ASTM D 772-86 Standard Test Method for Evaluating Degree of Flaking
Cure Time	D3732 Practice for Reporting Cure Times of Ultraviolet-Cured Coatings
Efflorescence	Efflorescence—ASTM D 7072-04 Standard Practice for Evaluating Accelerated Efflorescence
Gloss Evaluation	D4449 Test Method for Visual Evaluation of Gloss Differences between Surfaces of Similar Appearance
Viscosity and Flow	D4040 Test Method for Viscosity of Printing Inks and Vehicles by the Falling-Rod Viscometer
Volatile Content	D5403 Test Methods for Volatile Content of Radiation Curable Materials

Gaskets, Seals, and Belts

As discussed in Chapter 3, failure of these occurs for all the regular reasons that materials fail due to physical *change*:

- Deformation
- Fracture
- Wear
- Molecular transition

This *change* occurs because of the following environmental *influence*:

- Force
- Temperature
- Time
- Chemical

As with any piece of process equipment, to avoid recurring failure and downtime, it is imperative that the cause of the failure be addressed and not merely the symptoms. Scrutinizing the physical characteristics of failed seal faces and components will uncover the culprit and assist in determining the corrective actions to be taken. Gaskets, seals, and belts will experience the following opportunities for failure. Proper material selection is important to offset these modes:

Chemical Attack: This leaves the parts appearing dull, honeycombed, flaky, or starting to crumble or break up. Weight and material hardness readings taken on the damaged parts will be substantially lower than readings on the original parts.

Fretting Corrosion: This is one of the most common types of corrosion encountered, especially in seals. It only causes leakage at the secondary seals, but it damages the sleeve directly beneath the secondary seal area. This area will appear pitted and shiny bright.

Leaching: Normally causes a minor increase in seal leakage and a large increase in the wear of carbon faces. Ceramic and tungsten carbide faces that have been leached will appear dull and matted, even though no coating is present on them.

Erosion: The seal face may be eaten away or washed out in one localized area. Erosion will commonly occur on a stationary seal face until seal face distortion or breakdown occurs.

Heat Checking: This is indicated by the presence of fine to large cracks that seem to radiate from the center of the seal face. These cracks act as a series of cutting and scraping edges against carbon graphite and other seal face materials.

Vaporization: Any popping, puffing, or blowing of vapors at the seal faces is evidence of vaporization. Vaporization does not frequently cause catastrophic failure, but it usually shortens seal life. Inspection of the seal faces reveals signs of chipping at the inside and outside diameters and pitting over the entire area.

Oxidation: This leaves a varnish, a lacquer, or an abrasive sludge on the atmospheric side of the seal. This can cause rapid wear of the seal faces or hang-up in both pusher and nonpusher types of mechanical seals.

Gasket and Seal Properties

Gasket type
Gasket description
Cure method
Material
Gasket dimensions
Operating pressure (psi)
Operating temperature (°F)
Electromagnetic interference/radio-frequency interference (EMI/RFI) shielding

(*Continued*)

Gasket and Seal Properties (Continued)

Mechanical seal type
Seal design
Spring element
Specialized mounting
Seal configuration
Seal application
Location on shaft
Shaft diameter (inch)
Service limits
Rubbing speed (ft/min)
Shaft speed (rpm)
Operating pressure (psi)
Operating temperature (°F)
Direction of rotation
Face or primary ring materials
Mechanical carbon
Metal
Metal carbide and metal oxide (ceramic)
Plastics
Other face or primary seal material
Seat or mating ring material
Mechanical carbon
Metal
Metal carbide and metal oxide (ceramic)
Plastics
Other seat or mating ring material
Spring and metallic bellows materials

Gasket and Seal Properties (Continued)

Secondary seal or O-ring materials
Elastomers
Nonelastomers
Other secondary seal or O-ring material
Dynamic seal type
Seal orientation
Seal dimensions
Design units
Shaft outer diameter (O.D.) or seal inner diameter (I.D.) (inch)
Housing I.D. or seal O.D. (inch)
Axial X-section (inch)
Radial X-section (inch)
Material (ASTM abbreviation)
Service limits
Speed (ft/min)
Pressure (psi)
Vacuum
Operating temperature (°F)

Gasket and Seal Properties, Test Methods, and Standards

Property	*ASTM Test Methods and Standards*
Blowout Resistance	ASTM F434 - 93 Standard Test Method for Blow-Out Testing of Preformed Gaskets
Compression Resistance	F36 Test Method for Compressibility and Recovery of Gasket Materials
	D395 Test Methods for Rubber Property—Compression Set
Creep Resistance	F38 Test Methods for Creep Relaxation of a Gasket Material

(*Continued*)

Gasket and Seal Properties, Test Methods, and Standards (Continued)

Property	*ASTM Test Methods and Standards*
Flame Resistance	C1166 Test Method for Flame Propagation of Dense and Cellular Elastomeric Gaskets and Accessories
Flexibility	F147 Test Method for Flexibility of Nonmetallic Gasket Materials
Fluid Resistance	F146 Test Methods for Fluid Resistance of Gasket Materials
Fungus Resistance	G21 Practice for Determining Resistance of Synthetic Polymeric Materials to Fungi
Sealability	F37 Test Methods for Sealability of Gasket Materials
Temperature Resistance	D746 Test Method for Brittleness Temperature of Plastics and Elastomers by Impact
	D865 Test Method for Rubber-Deterioration by Heating in Air (Test Tube Enclosure)
Tensile Strength	F152 Test Methods for Tension Testing of Nonmetallic Gasket Materials
	D412 Test Methods for Vulcanized Rubber and Thermoplastic Elastomers—Tension
Thermal Conductivity	F433 Practice for Evaluating Thermal Conductivity of Gasket Materials
Water Resistance	C1083 Test Method for Water Absorption of Cellular Elastomeric Gaskets and Sealing Materials

Belt Properties

Belts
Flat Belt Physical Properties
Thickness (inch)
Width (inch)
Length (inch)
Weight PIW (lbs)
Working load PIW (lbs)
Maximum operating speed (ft/min)

Belt Properties (Continued)

Minimum pulley size (inch)
Operating temperature (°F)
Application
Material
V Belt Specifications
English
Metric
V-Ribbed
Belt angle (degrees)
Belt width (inch)
Belt length (inch)
Bands
Type
Number
Material
Belt
Reinforcement
Operating Specifications
Strength (lbs)
Maximum speed (rpm)
Minimum pulley size (inch)
Features
Operating temperature (°F)
Timing Belt Specification
Geometry
Pitch

(*Continued*)

Belt Properties (Continued)

Timing Belt Specification
Belt width (mm)
Number of teeth
Belt length (mm)
Material
Belt
Reinforcement
Performance Specifications
Ultimate tensile strength (N)
Maximum operating speed (rpm)
Minimum pulley size (grooves)
Operating temperature (°C)

Belt Properties, Test Methods, and Standards

Properties	*Belt Tests and Standards*
Abrasion Resistance	ASTM D1630 Test Method for Rubber Property-Abrasion Resistance (Footwear Abrader)
	ASTM D2228 Test Method for Rubber Property-Relative Abrasion Resistance by Pico Abrader Method
	ASTM D5963 Test Method for Rubber Property-Abrasion Resistance (Rotary Drum Abrader)
	DIN 53516 Testing of Rubber and Elastomers; Determination of Abrasion Resistance
	ISO 4649 Rubber-Determination of Abrasion Resistance Using a Rotating Cylindrical Drum Device
Electrical Conductivity	ISO 284 Conveyor Belts—Electrical Conductivity—Specification and Method of Test
Flame Resistance	30 CFR 18.65 Flame Testing of Conveyor Belting and Hose
	ISO 340 Conveyor Belts—Flame Retardation—Specifications and Test Method

Belt Properties, Test Methods, and Standards (Continued)

Properties	*Belt Tests and Standards*
Fluid Resistance	ASTM D471 Test Method for Rubber Property—Effect of Liquids
Hardness	ASTM D1415 Test Method for Rubber Property—International Hardness
	ASTM D2240 Test Method for Rubber Property—Durometer Hardness
Heat Resistance	ASTM D573 Test Method for Rubber—Deterioration in an Air Oven
	ASTM D865 Test Method for Rubber—Deterioration by Heating in Air (Test Tube Enclosure)
Resistance to Cracking	ASTM D1149 Test Methods for Rubber Deterioration—Cracking in an Ozone Controlled Environment
Slip Resistance	ASTM D1894 Test Method for Static and Kinetic Coefficients of Friction of Plastic Film and Sheeting
Tear Strength	ASTM D624 Test Method for Tear Strength of Conventional Vulcanized Rubber and Thermoplastic Elastomers
Tensile Strength	ISO 283 Conveyor Belts—Full Thickness Tensile Strength and Elongation—Specifications and Method of Test
	ASTM D412 Test Methods for Vulcanized Rubber and Thermoplastic Elastomers—Tension

Chains

Chains share the same failure modes as other components. Provided that the chain has been properly installed and sized accordingly, the typical reason for failure is due to wear, load, or corrosion. Wear of the plates and pins of bushings can occur due to the design of the machine or from abrasive debris. Material selection of the plates and pins can help offset corrosion that may be due to acid or caustic materials. Plastic chain components may offer a better alternative than steel.

Graph 4.1 represents a typical wear profile for chain. From point O to A, the chain is experiencing initial wear. This phase is short but intense. The elongation experienced versus time may be dramatic, but it is short-lived. After the initial wear phase, there is a longer duration of normal wear (A to B) that accounts for the

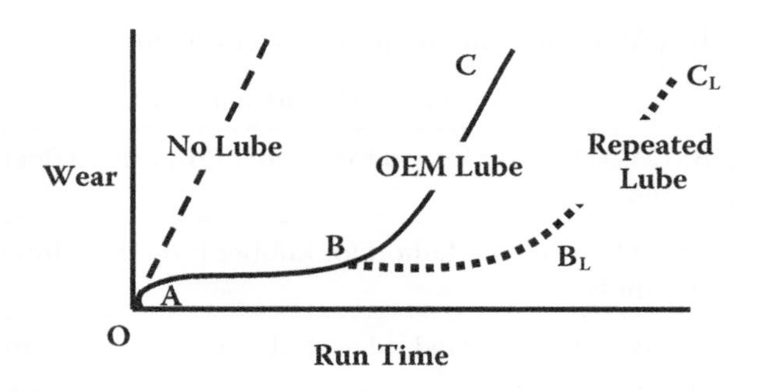

Graph 4.1 Chain life.

chain's normal wear pattern. After this phase, there is considerable increase in elongation and dramatic wear ensues (B to C). This phase is considered extreme wear. The values obtained during this stage of the chain's life determine the allowable wear limits. At this point, the chain has experienced its useful life span and is destined to fail or be replaced.

Graph 4.1 represents a chain that was lubricated when installed but not lubricated again. A chain that has been lubricated after installation will experience a much longer A to B zone. Removing the chain lubricant prior to installation will shorten the chain life dramatically, as seen by the dashed line.

Load failure is directly related to the tensile strength of the chain as well as the elongation percentage. The tensile loading of a chain is a measure of how well it behaves under load.

Elongation is typically the main reason why chains fail. As a load increases, the chain links become longer.

Chain Physical Properties

Chain
Roller Chain Specifications
Pitch
ANSI
BS/DIN
Strand Configuration
Number
Nonstandard number

Chain Physical Properties (Continued)

Chain Construction
Material
Options
Attachment
Specialty
Operating Specifications
Tensile strength (lbs)
Maximum allowable load (lbs)
Chain weight (lbs/ft)
Operating temperature (°F)

Chain Properties, Test Methods, and Standards

Standards for Major Types of Chains			
Chain Category	*ANSI Standard*	*ISO Standard*	*Japanese Industrial Standards (JIS)*
Power Transmission Roller Chain	ANSI B 29.1M	ISO 606	JIS B 1801
Power Transmission Bushed Chain	ANSI B 29.1M	ISO 1395	JIS B 1801
Power Transmission Sprocket	ANSI B 29.1M	ISO 606	JIS B 1802
Heavy-Duty Chain	ANSI B 29.1M	ISO 3512	
Bicycle Chain		ISO 9633	JIS D 9417
Motorcycle Chain		ISO 10190	JCAS 1[2]
Leaf Chain	ANSI B 29.8M	ISO 4347	JIS B 1804
Double-Pitch Conveyor Chain and Sprocket	ANSI B 29.4	ISO 1275	JIS B 1803
Power Transmission Roller Chain with Attachment	ANSI B 29.5		JIS B 1801
Conveyor Chain	ANSI B 29.15	ISO 1977/1~3	JCAS 2[2]

Wire Rope

There are several physical properties that will influence the useful life span of the wire rope or lead to failure.

Tensile Strength: The greater the load, the greater is the ability of the wire rope to withstand breaking. At some point, the load becomes greater than the material can withstand, and failure (strand breaking) ensues.

Fatigue Resistance: As the wire rope flexes, the wires become strained. Repeated flex will cause the wire to eventually break. The rapid bending of the wire rope can break individual wires in the strands. Many applications provide a constant bending and flexing of wire rope. Fatigue resistance is an important property when run at high speeds. Lang lay ropes are best for service requiring high fatigue resistance. Ropes with similar wires around the outside of their strands also have greater resistance, because these strands are more flexible.

Crushing Strength: This is the strength necessary to resist the compressive and squeezing forces that distort the cross section of a wire rope under a heavy load.

Abrasion Resistance: As wire rope flexes, the strands rub on each other. As the metal rubs on metal, the threads become smaller, wear particles beget more wear particles, and soon the wires fray and wear out. Wire rope must have the ability to withstand the gradual wearing away of the outer metal, as the rope runs across sheaves and hoist drums. Speed on load influences the rate of abrasion, but the type of metal and the size of the strands have a direct relationship on the longevity of the wire rope for a given application. Wire rope made of harder steels, such as improved plow steel, has a considerable resistance to abrasion. Ropes that have larger wires forming the outside of their strands are more resistant to wear than rope having smaller wires that wear away more quickly. The abrasion of wire rope can be reduced by using a wire rope lubricant that has penetrating ability and antiwear properties.

Rust and Corrosion Resistance: When wire rope is exposed to water or corrosive material, iron-based metal rusts and non-iron-based metal corrodes. Wire ropes can be protected from rust and corrosion by galvanizing, using various coatings or special lubricants. Wire rope used in crane operations should use special lubricants that also incorporate corrosion protection.

Wire Rope Failure: Some of the common causes of wire rope failure are the following:

- Using incorrect size, construction, or grade
- Subjecting to severe or continuing overload
- Using an excessive fleet angle
- Dragging over obstacles
- Lubricating improperly
- Operating over sheaves and drums of inadequate size

- Overriding or cross winding on drums
- Operating over sheaves and drums with improperly fitted grooves or broken flanges
- Jumping of sheaves
- Exposing to acid or corrosive liquid or gases
- Using an improperly attached fitting
- Allowing grit to penetrate between the strands, promoting internal wear

Wire Rope Physical Properties

Specifications
Strength (lbs)
Diameter (inch)
Length (inch)
Construction
#Strands X #Wires
Strands
Wires/strand
Type
End fittings
Material

Wire Rope Properties, Test Methods, and Standards

Properties	*Test Methods and Standards*
Carbon Steel General Specification	ASTM A1023/A1023M-07 Standard Specification for Stranded Carbon Steel Wire Ropes for General Purposes
	AIME/ISS Carbon Steel, Wire, and Rods
General Wire Rope Specification	API 9A Specification for Wire Rope
	EN 10264-1.2 Steel Wire and Wire Products—Steel Wire for Wire Rope
	ISO Std. 2232 Drawn Wire for General Purpose Nonalloy Steel Wire Ropes

(*Continued*)

Wire Rope Properties, Test Methods, and Standards (Continued)

Properties	*Test Methods and Standards*
Tensile Strength	ASTM A931-08 Standard Test Method for Tension Testing of Wire Ropes and Strand
	E8 Test Methods for Tension Testing of Metallic Materials
Torsion Strength	A938 Test Method for Torsion Testing of Wire
Zinc-Coated Steel General Specification	ASTM A603-98(2003) Standard Specification for Zinc-Coated Steel Structural Wire Rope

Pipe

Pipe can fail for many reasons:

- Material degradation
- Incorrect pipe material selection
- Bursting
- Crushing
- Abrasion
- Improper fitting application
- Improper adhesive bonding

Failure of the Fitting

Transition: The plastic-to-steel transition fitting fails at the mechanical pullout-proof metal and plastic couplings and factory produced weld-in and fuse-in transition fittings.

Valve (plastic): Fusion or mechanically joined types.

Meter Risers (two types): (1) Anodeless—The plastic is the gas carrier up inside the steel riser casing to a point above grade. All risers installed today are the "anodeless" type. (2) Steel riser—The plastic is fitted to a steel compression fitting, below ground, at the end of the horizontal leg of the riser.

Mechanical Fitting: The failure occurred within the body of a mechanical fitting. This includes stab type, screw on, bolt on, and mechanical tapping tees made of metal and plastic. This will include both pullout-proof and seal-only types but does not include plastic-to-steel couplings—they are considered transition fittings. This would apply only to failures in the bodies of mechanical fittings and not to failures in the joints between the fittings and pipe.

Heat Fusion: The failure occurred in a conventional or hot plate fusion fitting. This would include conventional plastic fusion fittings such as socket fusion couplings, "L"s and "T"s, and saddle fusion tees. This would not include butt fusion joints but would include failures in the mold seams of molded plastic fittings. This would only apply to failures in the bodies of fusion fittings and not failures in the joints between the fittings and pipe.

Electrofusion: Includes electrofusion saddles, patching saddles and couplings, "L"s, and "T"s. This failure mode also includes failures in the bodies or seams of molded, extruded body electrofusion fittings. This would not apply to failures in the joints between the fittings and pipe or butt fusion joints.

Joint Failure

There are a variety of ways in which pipes are joined:

Mechanical: A joint where pipes are joined by bolting or threading their ends together.

Adhesive/Solvent: Solvents are used to join polyvinyl chloride (PVC) drain, waste, vent (DWV) pipe. The solvent is used to soften and "glue" two pipe sections together.

Welded: Metal and some plastic pipes can be welded. Plastic pipe uses a hot plate to melt the ends of the plates to be joined. The plate is removed, and the ends are pushed together using joining machinery, creating a seamless joint.

Bell and Spigot: Bell and spigot joints are often used in gravity lines. With bell and spigot joints, each pipe length has a bell (or larger diameter end piece) end and spigot (or normal diameter) end. The spigot is inserted into the bell via a compression fit. Much sewer work uses bell and spigot joints.

The failure was in the joint between the pipe and the fitting or between two sections of pipe.

Adhesive Failure: Failure occurred in the joining area of a solvent cemented joint. This would only apply to solvent cement plastics such as PVC or acrylonitrile butadiene styrene (ABS). The solvent used may have created a weakened interface resulting in failure (this may also apply to the solvent in the adhesive) or the adhesive may not provide adequate strength or it may have compromised the physical integrity of the material sought to adhere.

Excessive Expansion/Contraction: The failure could be due to the pipes' ductile strength. The failure is often due to third-party damage caused by a backhoe or excavator. An easy check for this sort of failure can be done by examining the pipe for signs of scraping where it was inserted into the coupling; the failure is likely to be due to poor installation, especially if the coupling is a pull-out-proof type. If the failure is slow crack growth adjacent to the fitting, this could be due to fatigue or excessive bending caused by point loading at the joint or a failure to properly sleeve or obtain support through backfilling.

Excess External Earth Loading: Examination of the failed section of pipe will reveal if the pipe was excessively bent, kinked, or mishandled. If the failure is the brittle type, it will occur at the area of maximum stress. If this is where a protective sleeve should have been installed, or where a moving load was being applied, such as a driveway, and the pipe installed was shallow or poorly backfilled, then the cause is clear. Examination of the leak and the area of the leak should provide the answer. This type of failure usually occurs because of bending and stress loading where a pipe (more flexible) enters a fitting such as a socket fusion fitting (less flexible).

Fusion Failures: Occur in joints as well as fittings (discussed earlier):

- Electrofusion—The failure occurs in the electrofusion joint. This would include couplers, "L"s, "T"s, saddle "T"s, branch saddles, and patching saddles.
- Butt Fusion—The failure occurred in a hot-plate-fused butt fusion joint.
- Socket Fusion—The failure occurred in a hot-plate-fused socket fusion joint.
- Saddle Fusion—The failure occurred in a hot-plate-fused saddle fusion joint. This would not include electrofusion saddles or mechanical saddles.

Installation Error: Leakage caused by a failure to follow proper installation procedures or operating instructions.

Mechanical Joint Failure: Failure occurs for the same reasons all materials would experience failure: exceeding the tensile strength, elongation percentage, or impact strength of the material of construction.

Point Loading: There is evidence of a foreign object (e.g., rock, tree root, etc.) pushing or rubbing against the pipe. (This is one cause of a brittle-type crack.)

Previous Impact: A good example would be a brittle crack through the pipe wall next to a third-party (outside force) gouge that initiated the crack in the pipe wall. If the pipe failure occurred sometime after the impact, then this is reported as a Previous Impact pipe failure using the plastic pipe database form. Failures reported or eligible to be reported as third-party damage under the Common Ground Initiative would not be reported again under this category.

Pipe Physical Properties

Dimensions
Diameter
Length (ft)
Wall thickness (inch)
Schedule

Pipe Physical Properties (Continued)

Dimensions
Shape
Ratings
Maximum pressure (psi)
Temperature range (°F)
Weight (lb/foot)
Features
Material
Connections
Color

Pipe Properties, Test Methods, and Standards

Properties	*Test Methods and ASTM Standards*
Chemical Resistance of Plastic Pipe	D543 Practices for Evaluating the Resistance of Plastics to Chemical Reagents
Crack Growth Resistance of Polyethylene (PE) pipe	F1473 Test Method for Notch Tensile Test to Measure the Resistance to Slow Crack Growth of Polyethylene Pipes and Resins
Electrical Arc Resistance	A513 Specification for Electric-Resistance Welded Carbon and Alloy Steel Mechanical Tubing
	A587 Specification for Electric-Resistance Welded Low-Carbon Steel Pipe for the Chemical Industry
General Specification for Carbon Steel Castings	A234/A234M Specification for Piping Fittings of Wrought Carbon Steel and Alloy Steel for Moderate and High Temperature Services
General Specification for Carbon Steel Pipe	A105/A105M Specification for Carbon Steel Forgings for Piping Applications
General Specification for Cast Iron Fittings	B16.1 Cast Iron Pipe Flanges Flanged Fittings

(*Continued*)

Pipe Properties, Test Methods, and Standards (Continued)

Properties	*Test Methods and ASTM Standards*
General Specification for Ductile Iron	A536 Specification for Ductile Iron Castings
General Specification for Ductile Iron Fittings for Pressure	B16.42 Ductile Iron Pipe Flanges and Flanged Fittings—Section IX of the ASME Boiler and Pressure Vessel Code
General Specification for Ferritic Ductile Iron Castings for High Temperature	A395/A395M Specification for Ferritic Ductile Iron Pressure-Retaining Castings for Use at Elevated Temperatures
General Specification for Gray Iron Castings	A126 Specification for Gray Iron Castings for Valves, Flanges, and Pipe Fittings
	A48/A48M Specification for Gray Iron Castings
General Specification for Pressure on Gray Iron Castings	A278/A278M Specification for Gray Iron Castings for Pressure-Containing Parts for Temperatures Up to 650°F (350°C)
General Specification for Stainless Steel Pipe—High Temperature	A106 Specification for Seamless Carbon Steel Pipe for High-Temperature Service
General Specification for Steel Castings	A182/A182M Specification for Forged or Rolled Alloy-Steel Pipe Flanges, Forged Fittings, and Valves and Parts for High-Temperature Service
	A216/A216M Specification for Steel Castings, Carbon, Suitable for Fusion Welding, for High-Temperature Service
General Specification for Steel Fittings	B16.5 Steel Pipe Flanges and Flanged Fittings
General Specification for Wrought Steel Elbows and Returns	B16.28 Wrought Steel Buttwelding Short Radius Elbows and Returns
General Specification for Wrought Steel Fittings	B16.9 Factory-Made Wrought Steel Butt Welding Fittings
	MSS SP-43 Wrought Stainless Steel Butt-Welding Fittings

Pipe Properties, Test Methods, and Standards (Continued)

Properties	*Test Methods and ASTM Standards*
General Specification for Zinc Coated	A53/A53M Specification for Pipe, Steel, Black and Hot-Dipped, Zinc-Coated Welded and Seamless
General Specification of PE Pipe Fittings	D3350 Specification for Polyethylene Plastics Pipe and Fittings Materials
General Specification of PVC Pipe Fittings	D2466 Specification for Poly(Vinyl Chloride) (PVC) Plastic Pipe Fittings, Schedule 40
	D2467 Specification for Poly(Vinyl Chloride) (PVC) Plastic Pipe Fittings, Schedule 80
	D2564 Specification for Solvent Cements for Poly(Vinyl Chloride) (PVC) Plastic Piping Systems
	D2672 Specification for Joints for IPS PVC Pipe Using Solvent Cement
	D2740 Specification for Polyvinyl Chloride (PVC) Plastic Tubing
Impact Resistance of Plastic Pipe	D2444 Test Method for Determination of the Impact Resistance of Thermoplastic Pipe and Fittings by Means of a Tup (Falling Weight)
Load Strength of Plastic Pipe	D2412 Test Method for Determination of External Loading Characteristics of Plastic Pipe by Parallel-Plate Loading
Pressure Resistance for Plastic Pipe	D1598 Test Method for Time-to-Failure of Plastic Pipe Under Constant Internal Pressure
Pressure Resistance for PVC and Standard Dimension Ratio (SDR) Pipe	D2241 Specification for Poly(Vinyl Chloride) (PVC) Pressure-Rated Pipe (SDR Series)
Pressure Resistance of Plastic Pipe for Hydraulics	D1599 Test Method for Resistance to Short-Time Hydraulic Pressure of Plastic Pipe, Tubing, and Fittings

(*Continued*)

Pipe Properties, Test Methods, and Standards (Continued)

Properties	*Test Methods and ASTM Standards*
Tensile Strength for Plastic Pipe	D638 Test Method for Tensile Properties of Plastics
	ISO 13953 Polyethylene (PE) Pipes and Fittings—Determination of the Tensile Strength and Failure Mode of Test Pieces from a Butt-Fused Joint
	D2290 Test Method for Apparent Hoop Tensile Strength of Plastic or Reinforced Plastic Pipe by Split Disk Method
	D638 Test Method for Tensile Properties of Plastics
Thermoplastic Gas Pressure Resistance	D2513 Specification for Thermoplastic Gas Pressure Pipe, Tubing, and Fittings

Hose

Depending on the material and reinforcement, hose can stretch up to 2% and contract as much as 4% in length and girth. This can strain hose reinforcement wires and coupling interfaces, leading to failure. Cutting the hose to compensate is important. Another important consideration to reduce failure is reducing hose twist. If hose is twisted 5°, this can reduce service life by 70%. A 7° twist can reduce service life up to 90%. Often the routing of hose is one of the last tasks and is subject to a rather arduous path.

It is often suggested by hose manufacturers to use a single section of hose for each bend and install a hose-to-hose coupling and hose clamp between bends. This technique is less preferred, because it not only is more costly and time consuming to perform, but it increases the number of potential leak points in the hose assembly. Also, to help ensure that technicians replace and secure hose assemblies properly, include detailed instructions on hose length, use of hose clamps, and special considerations in service manuals. The best course of action is to plan the path of the hose in advance, cut to compensate, and use couplings and fittings sparingly.

Another source of failure comes from heat. Heat from external sources, such as exhaust systems, motors, or ovens, can first soften hose by bringing the polymer to its melting point. Extended exposure just under the melting point will drive out the plasticizing oils used in the polymer matrix, leaving the hose brittle and susceptible to bursting. If the hose is for hydraulic applications, a high heat source can melt the hose, if the pressurized stream of hydraulic fluid comes in contact with the heat source and ignites, it is possible that what is created is nothing short of a flame thrower. Once this occurs, there is a very good chance that anything in the surrounding area will catch fire. If hose is going to be near a heat source, it should be insulated with protective sleeves to partially block heat transmitted to the hose.

Heat from internal sources can also be an issue. In hydraulic applications, the sump for the oil must be adequate in size to accommodate enough fluid to act as a heat sink. The hydraulic fluid can also reduce the service life of the hose. Pumping fluid can contribute to internal heat buildup, especially if the fluid is pressurized. Fluids will not compress, but the force of pressure will restrict flow which will raise the internal temperature. The hose lining is susceptible to softening if the fluid is not cooled.

Physical damage accounts for over 80% of hose failures. The physical damage most noted is abrasion primarily generated by repeatedly rubbing against equipment surfaces or each other. Using clamps to secure hose will reduce abrasion. Protection can also be provided by using sleeves. Metal sleeves that look like springs will protect the hose from abrasion and crushing and will expand and contract with the hose while remaining flexible. If abrasive particulates are an issue, then a reinforced fabric sleeve may be in order. Sleeves are also used to bundle hoses.

Hose Physical Properties

Product Form
I.D. (inch)
O.D. (inch)
Weight (lbs/ft)
Minimum bend radius (inch)
Working pressure (psi)
Burst pressure (psi)
Maximum vacuum (in Hg vac)
Temperature range (°F)
Material
Specialized construction
Features
Media
Application
Color
Length

Hose Properties, Test Methods, and Standards

Properties	*ASTM Test Methods and Standards*
Electrical Arc Resistance	D2865 Practice for Calibration of Standards and Equipment for Electrical Insulating Materials Testing
Fluid Resistance	D471 Test Method for Rubber Property—Effect of Liquids
Hardness	D1415 Test Method for Rubber Property—International Hardness
	D2240 Test Method for Rubber Property—Durometer Hardness
Heat Resistance	D572 Test Method for Rubber—Deterioration by Heat and Oxygen
	D573 Test Method for Rubber—Deterioration in an Air Oven
	D865 Test Method for Rubber—Deterioration by Heating in Air (Test Tube Enclosure)
Ozone Resistance	ASTM D 1149 Rubber Deterioration—Cracking in an Ozone Controlled Environment
Tear Strength	D624 Test Method for Tear Strength of Conventional Vulcanized Rubber and Thermoplastic Elastomers
Tensile Strength	D412 Test Methods for Vulcanized Rubber and Thermoplastic Elastomers—Tension
Water Resistance	D570 Test Method for Water Absorption of Plastics

Valves

Valves will fail for a number of reasons, but the most common are due to wear and corrosion. Pressure and bursting are typically not the reasons for valve failures. The flow rates and operating pressures are well understood and calculated. Over time, the turbulent flow of material can slowly and steadily wear the internal components down. When this occurs, flow rate and volume discrepancies occur. This situation may be disastrous for the cooling of a heating system or for a specific amount of product that has to be applied precisely into a product. Improper amounts can result in additional scrap to catastrophic failures resulting in loss of life. There are several types of wear that can and do take place on valves: adhesive, abrasive, and corrosive.

Adhesive Wear (two-part wear): As mentioned, adhesive wear occurs at the interface where metal contacts metal. When this occurs, pressures and temperatures increase and this facilitates microscopic welding. The heat produced at the contact interface is very high, near the melting point of the two metals. The tiny points of metal sticking up, also known as asperities, contact similar points on the other part, and as seating of the valve continues, these points come under very high stress because they are initially carrying the entire load. The rubbing of these points over one another is what creates the high surface temperature. Also, each time the valve seats, it has moved slightly, so no new points, or asperities of the mating parts, are touching each other. That roughened surface is the result of the microscopic welds being torn apart as the valve and seat separate.

Abrasive Wear (three-part wear): As mentioned, abrasive wear is wear that is caused by an abrasive substance found between two components. The substance (normally only several microns in size) digs into the surface of the metal, producing stress fractures and breaking other small pieces away for the component. These other pieces, in turn, continue to ravage the component until there is a significantly dimensional change as well as a change in the finish. Abrasive wear can also occur with nonparticles. Fluids can produce abrasive wear—case in point, the Grand Canyon. The Colorado River carved out the great gorge. Some would suggest that the river brought through silt, making the actual mechanism of wear achieved by slurry and not a liquid. If that be the case, then there are countless examples of wear created by water alone along the Eastern seaboard of the United States.

Corrosive/Chemical Wear: Metal and plastic can be sensitive to the effects of various compounds they come in contact with. Water as well as caustics and acids will attack most materials in short order. It is important to review the material makeup of the components and compare the corrosion-resistance information to the material that is to be pumped.

Valve Physical Properties

Valve Physical Properties
Valve type
Primary material
Valve size (inch)
Maximum pressure (psi)
Number of ports
Media temperature (°F)
Flow (Cv)
Actuation
Connection description

Valve Properties, Test Methods, and Standards

Properties	*Test Methods and Standards*
Low-Temperature Service for Forged Carbon and Alloy Steel	ASTM A 707/A 707M Specification for Forged Carbon and Alloy Steel Flanges for Low-Temperature Service
High-Temperature Service for Carbon Steel	ASTM A106 Specification for Seamless Carbon Steel Pipe for High-Temperature Service
High-Pressure Service for Metal-Arc Welded Steel Pipe	ASTM A381 Specification for Metal-Arc-Welded Steel Pipe for Use with High-Pressure Transmission Systems
General Specification for Steel Forgings	ASTM A788 Specification for Steel Forgings, General Requirements
	ASTM A961 Specification for Common Requirements for Steel Flanges, Forged Fittings, Valves, and Parts for Piping Applications
General Specification for Line Pipe	5L Specification for Line Pipe
General Specifications for Ferris Valves	ASME B 16.10 Face-to-Face and End-to-End Dimensions of Ferrous Valves
General Specification for Forged Steel Fittings	ASME B 16.11 Forged Steel Fittings, Socket Welding and Threaded
General Specification for Wrought Steel Elbows	ASME B 16.28 Wrought Steel Butt-Welding Short Radius Elbows
General Specification for Steel Flanges	ASME B 16.47 Large Diameter Steel Flanges
	ASME B 16.5 Steel Pipe Flanges and Flanged Fittings
	ASME B 16.9 Steel Butt-Welding Fittings
	MSS SP-44 Standard for Steel Pipe Line Flanges
General Specification for Swage Fittings	MSS SP-95 Swage (d) Nipples and Bull Plugs

Fasteners—Screws, Bolts, Nails, Staples

Fasteners experience static loading or fatigue loading. Static loading is the tension, shear, bending, or torsion that can occur in combination. Fatigue loading is the result of vibration or subtle stresses. In addition to static load and fatigue, some other common reasons for fastener failures include corrosion, manufacturing irregularities, and faulty installation.

When engaged in a failure mode effect analysis (FMEA), it would be important to understand the following in order to implement corrective actions. The following are some common questions concerning fasteners:

- How were the fasteners torqued?
- In what order were fasteners tightened?
- What is the best way to verify the torque on fasteners?
- How does torque value vary over time?

Fretting normally results from small movements between the surfaces. This can be due to vibration or poor spacing. Material selection, heat treatment, cutting or rolling threads, manufacturing, assembly, and design are some of the factors that affect fastener failures. Failure analysis can determine the cause of the fastener failure and determine the primary or contributing causes of fastener failure. A fastener may fail at one of several locations. The frequent locations for fastener failure are as follows:

- On the head-to-shank fillet
- Through the first thread inside the nut
- At the transition from the thread to the shank

When a fastener fails, it is typically due to one of the following reasons:

- Shear
- Overload
- Fatigue
- Corrosion
- Manufacturing discrepancies
- Improper installation

There are instances that when failure occurs on a component, it may not be directly associated with fasteners, yet by virtue of proximity, the fastener is blamed.

Consider the fastener a victim of circumstance. The following situations were failures not directly due to the fastener:

- Delamination of material near the hole
- Differences in expansion coefficients between mating surfaces and the fastener
- Drilling damage
- Fretting
- Leaks around the fastener
- Galvanic corrosion between surface and the fastener
- Installation damage
- Pullout of the fastener under load

When failure with a threaded fastener occurs, it generally occurs at one of three modalities: shank, external threads, or internal thread stripping.

Shank failure occurs due to torsional stress. The thread strength is determined by the stress area. This is based on the averaged diameter. The *Machinery's Handbook* (Industrial Press, 2007) is a great source of information on the stress areas of various thread sizes.

When a bolt is tightened, the shank undergoes both direct stress due to the elongation strain and torsion stress due to the torque acting on the threads. High-performance bolts are often designed so that the plain shank is smaller than the stress diameter of the thread. This is done so that the stretch that occurs under the preload induced from the tightening process is maximized. With this type of bolt, failure, if overtightened, will occur in the plain shank region. The thread stripping undergoes a shearing action that produces thread bending due to high loads and nut dilation. The following scenarios have a direct effect on fastener failures:

- Thread dimension variations—diameters and pitch.
- Tensile and shear strength variations in the material for both the internal and external threads.
- Radial displacement of the nut or tapped component (generally known as nut dilation) in reducing the shear strength of the threads—the tensile force in the fastener acts on the threads, and a wedging action generates a radial displacement that reduces thread strength.
- Bending of the threads, caused by the action of the fastener's tensile force acting on the "V" threads, resulting in a wedging action that decreases the shear area of the threads.
- Production variations in the threaded assembly, such as slight hole taper or bell mouthing, that can affect thread strength.

Holes exhibiting bell mouthing will, when tapped, experience a variable thread height along the length of the hole. This variation can be significant on short lengths of engagement and fine pitches. The net effect of bell mouthing is to reduce the shear area of the external threads. The finer the thread, the more pronounced is the effect of bell mouthing.

Fastener Physical Properties

Washers
Material
Diameter—inside
Diameter—outside
Plating
Thickness
Thread/screw/hole size
Type
Bolt Type
Length (inch)
Other size specifications
Material
Finish
Screw Type
Length (inch)
Other size specifications
Head type
Standard drives
Material
Finish
Rivet Type
Head type
Body/shank diameter (inch)
Grip range/length (inch)
Head diameter (inch)

(*Continued*)

Fastener Physical Properties (Continued)

Rivet Type
Head height (inch)
Shear strength (lbs)
Material
Finish
Nut Type
Units
Direction
Material
Finish

Fastener Properties, Test Methods, and Standards

Properties	*Test Methods and ASTM Standards*
Bend Strength of Nails	F1575 Test Method for Determining Bending Yield Moment of Nails
Carbonization	F2328M Test Method for Determining Decarburization and Carburization in Hardened and Tempered Threaded Steel Bolts, Screws, and Studs (Metric)
Corrosion Resistance	Galvanic Corrosion between Surface and the Fastener
Embrittlement Resistance	F1624 Test Method for Measurement of Hydrogen Embrittlement Threshold in Steel by the Incremental Step Loading Technique
General Specification for Fasteners	ASTM F606-07 Standard Test Methods for Determining the Mechanical Properties of Externally and Internally Threaded Fasteners, Washers, Direct Tension Indicators, and Rivets
General Specification for Nails, Spikes, and Staples	ASTM F1667-05 Standard Specification for Driven Fasteners: Nails, Spikes, and Staples

Fastener Properties, Test Methods, and Standards (Continued)

Properties	*Test Methods and ASTM Standards*
General Specification for Rivets	ASTM A31-04e1 Standard Specification for Steel Rivets and Bars for Rivets, Pressure Vessels
General Specification for Compression Washers	F959M Specification for Compressible-Washer-Type Direct Tension Indicators for Use with Structural Fasteners (Metric)
General Specification for Nails	F680 Test Methods for Nails
General Specification for Screws	F835M Specification for Alloy Steel Socket Button and Flat Countersunk Head Cap Screws (Metric)
General Specification for Steel Washers	F436M Specification for Hardened Steel Washers (Metric)
General Specification for Screws	F912M Specification for Alloy Steel Socket Set Screws (Metric)
Hardness	E10 Test Method for Brinell Hardness of Metallic Materials
	E18 Test Methods for Rockwell Hardness of Metallic Materials
	E92 Test Method for Vickers Hardness of Metallic Materials
	E384 Test Method for Microindentation Hardness of Materials
Shear Strength	B769 Test Method for Shear Testing of Aluminum Alloys
Tensile Strength	E8M Test Methods for Tension Testing of Metallic Materials (Metric)
	F606 Test Methods for Determining the Mechanical Properties of Externally and Internally Threaded Fasteners, Washers, Direct Tension Indicators, and Rivets

Bearings

There are many reasons why bearings will fail. Unfortunately, prior to failure, undue demands are put on motors which also lead to their eventual failure. The moving elements (rollers or balls) are points most susceptible. Often, high-temperature alloys are used for bearing fabrication, yet the mitigating factor for failure rests on the lubricant's shoulders (grease or oil). Bearing load ratings are determined using fatigue tests as well as temperature and stress load. In the field, conditions that contribute to bearing failures would be:

Misalignment: Not unlike when automobiles tires are misaligned, stress is not evenly applied, resulting in disharmonious vibrations that will lead to premature wear.

Vibration: Resulting from misalignment or from an external source. Low or high vibration can wear away at a rolling element or raceway. Often, bearing manufacturers will recommend that bearings stored on shelves be rotated and turned every month. This is due to the low-frequency vibrations of the plant resonating through the floor. Imagine the vibration on a piece of equipment.

Severe Loading: Sometimes due to the bearing design and sometimes due to the grease's inability to absorb shock.

Contamination: Sometimes due to the bearing design (sans a shield) and sometimes due to the grease's inability to resist water washout.

Severe Temperatures: Temperatures above 350°F (when cooled the temper and harness change) or below –20°F (many alloys become brittle) will adversely affect most of the common alloys used in bearings.

Too Much or Too Little Lubrication: Too much lubrication restricts movement and can raise operation temperatures, and too little provides little protection from metal-to-metal contact.

Poor Performing, Poor Quality Lubrication: See the "Grease" section below.

Improper Installation: Some technicians have the opinion that a dead blow sledge is a chosen method of application. Unless a tool designed to install bearings is used, the risk of applying localized stress incidents will occur, resulting in bearing failure.

Wrong Application: Wrong bearing.

The most common failure modes associated with bearing failure are:

- Load
- Wear
- Cold welding (seizing)
- Fretting

- Brinneling
- Corrosion
- Electrical arcing

It must be realized that the load applied to a bearing is enormous. Consider that the full load is focused on a very small surface area of the rolling element and the raceway of the bearing. If one was to calculate the pounds per square inch, it would easily exceed 250,000 PSI regardless of the bearing application, size, or even design.

Bearing Physical Properties

Bearing design
Type
Description
Design units
Bore (inch)
Outside diameter (inch)
Outer ring width (inch)
Overall width (inch)
Load ratings
Static axial (lbs)
Static radial (lbs)
Dynamic axial (lbs)
Dynamic radial (lbs)
Speed ratings
Grease (rpm)
Oil (rpm)
Material/plating
Rolling element
Seals/shields
Lubrication
Operating temperature (°F)

Grease

Grease failure can occur for several reasons, and it can be always traced back to lack of performance of the thickener, base oil(s), or performance additives.

Shearing: As grease is churned in a bearing, it can break down and lose consistency. When the grease loses its body, it liquefies and can drip out of the bearing, resulting in failure.

Oxidation: When heat and shear stress build, the base oil of the grease can begin to break down and oxidize. When this happens, a varnish layer forms on the rolling elements as well as the static components. The buildup of varnish will change the bearing tolerance and contribute to a temperature increase, as will the insulating varnish layer.

Water Washout: Water, steam, and even condensation can wash grease out of the bearing, leaving it unprotected. The grease serves not only as a lubricating barrier but also as a rust and corrosion inhibitor. If the grease becomes soluble, it will no longer afford protection.

Extreme Pressure and Wear: Point contact, similar to ball bearings or line contact, similar to roller bearings. This occurs because focused pressure breaks the grease film, allowing metal-to-metal contact resulting in component wear and eventual failure.

Rust/Corrosion: Breakdown of the alloy to form a useless alloy. Corrosion occurs due to lack of protection from the grease.

Grease Physical Properties

NLGI number
Low-temperature pumpability
Dropping point
Temperature range (°F)
Load-carrying ability
Mechanical stability
Wear resistance
Water resistance
Corrosion resistance
Oxidation resistance

Grease Properties, Test Methods, and Standards

Properties	*Grease Test Methods and Standards*
Shear Stability	ASTM D 217 Multistroke Penetration: Test measures the percent change in viscosity of the grease between an unworked and a worked state. The lower the percent change, the more mechanically stable the grease.
	ASTM D 1263 Wheel Bearing Leakage: Measures the percent loss in a wheel bearing application. The lower the percentage, the better; above 5% will cause brake problems.
Oxidation Resistance	ASTM D 942 Bomb Oxidation: Measures the oxidative life of the grease; this is used to help determine the shelf life.
Water Resistance	ASTM D 1264 Water Washout: Measures the resistance of grease to washout; the lower the percent, the less likely it will washout.
	ASTM D 4049 Water Spray-Off: Measures the resistance of a grease to spray-off; the lower the percent, the less likely it will wash out.
Bleed Resistance	FTM 321.3 Oil Separation (Static): Measures the percent oil that may separate during storage and idle time; the lower the percent, the more resistant the grease is to separating.
	ASTM D 1742 Pressure Oil Separation: Measures the percent oil that will separate when grease is under load; the lower the percent, the more resistant the grease is to separating.
Extreme Pressure/ Antiwear	ASTM D 2596 Four Ball: Measures point contact, similar to ball bearings; the higher the number, the greater load carrying ability of the grease.
	ASTM D 2266 Four Ball (Wear Scar): Measures wear protection the grease provides; the lower the number, the more protection the grease provides.
	ASTM D 2509 Timken Method: Measures line contact, similar to roller bearings; the higher the number, the greater load carrying ability of the grease.
Corrosion	ASTM D 1743 Rust Test: Is a static test that determines how well the grease keeps water and corrosives away from the metal surface.
	ASTM D 130 Copper Corrosion: Measures the ability of the grease to protect yellow metals.

(*Continued*)

Grease Properties, Test Methods, and Standards (Continued)

Properties	*Grease Test Methods and Standards*
Pumpability	US STEEL LT37 Mobility: Measures the grease flow at a given temperature at 150 psi; the higher the number, the better.
Service Temperature	ASTM D 2265 Dropping Point: Measures the temperature that the soap in the grease melts; this is used to help determine the upper operating temperature range.

Gears

The American Gear Manufacturers Association (AGMA) reports that the failure sequence of a gearbox frequently starts with a bearing failure, rather than a gear failure. If the source of failure is not bearing related, then it has been understood that gears can fail in several different ways. Excessive and even subtle vibration can facilitate gear failures. When gears mesh, they roll at the pitch line. Above and below this line, sliding action causes wear that can lead to failure. The AGMA recognizes four gear failure modes:

Wear: Occurs when metal is worn away from the contact areas of the gear teeth in a more or less uniform manner. Wear can occur due to a breakdown in lubricant performance (which places the reason for failure upon the gear lubricant) or from abrasive particles that generate wear particles that also contribute to gear wear.

Surface Fatigue: Metal removal and formation of cavities. These may be small or large and may grow or remain small. It occurs when the gear material fails after repeated stresses that are beyond the endurance limits of the metal.

Plastic Flow: Surface deformation that occurs when high-contact stresses combine with the rolling and sliding action of the meshing gear teeth to cause cold working of the tooth surfaces. Although usually associated with softer materials, it can also occur in heavily loaded case-hardened and through-hardened gears.

Breakage: Fracture of a whole tooth or substantial part of a tooth. Common causes include overload and cyclic stressing of the gear tooth material beyond its endurance limit.

Gears make contact in small areas or along narrow bands. Each part of the gear-tooth surface is only in contact for a short duration of time. Gear-tooth surface alignment affects the loading in use. The dynamic of failure is very pronounced due to the stress level associated with the sliding and rolling action of the gear face. If the tolerances are off just a few thousands of an inch, the difference can facilitate a quick failure. A buildup of varnish or contaminates, the misalignment of the shaft and bearings, and undue vibration can all lead to gearbox failure.

Gear Physical Properties

Gear type
Number of teeth
Design
Direction
Pressure angle
Gear center
Tooth form
English gear dimensions
Pitch
PD (inch)
O.D. (inch)
Face width (inch)
Overall width (inch)
Length (inch)
Bore diameter (inch)
Shaft diameter (inch)
Metric gear dimensions
Module
PD (mm)
O.D. (mm)
Face width (mm)
Overall width (mm)
Length (mm)
Bore diameter (mm)
Shaft diameter (mm)
Material
AGMA
Mounting

Gear Oil

The way in which gear oil fails is similar to the failure mechanism for grease in certain instances. Gear oils can fail for the following reasons:

Seal and Gasket Embrittlement: Certain gear oils will actually draw the plasticizing agents out of the rubber, rendering them weakened and susceptible to leaks.
Oxidation: When heat and shear stress build, the base oil of the grease can begin to break down and oxidize. When this happens, a varnish layer forms on the rolling elements as well as the static components. The buildup of varnish will change the bearing tolerance and contribute to a temperature increase as will the insulating varnish layer.
Foam: Considered unlubricated hot spots, foam can lead to increased oxidation as well as increased wear.
Water Washout: Water, steam, and even condensation can mix with gear oil. The gear oil serves not only as a lubricating barrier but also as a rust and corrosion inhibitor; therefore, if the gear oil becomes water soluble, it will no longer afford protection.
Mechanical Wear: Point contact occurs because focused pressure breaks to oil film allowing metal-to-metal contact resulting in component wear and eventual failure.
Rust/Corrosion: Breakdown of the alloy to form a useless alloy. Corrosion occurs due to lack of protection from the grease.
Oil Viscosity Builds or Thins: As oil thickens, reducing oil flow, oil thins out, allowing metal-to-metal contact.

Gear Oil Physical Properties

Kinematic viscosity (cSt)
Viscosity index
Specific gravity
Temperature range(°F)
Pour point (°F)
Thermal conductivity (W/m-K)
Flash point (°F)
Wear resistance
Water demulsibility
Oxidation resistance
Corrosion resistance

Gear Oil Properties, Test Methods, and Standards

Gear Oil Property	*Test Methods and Standards*
Gear Oil Operating Viscosity	ASTM D 2270: Viscosity Index: Measures variation in viscosity due to changes in temperature. The higher the number, the more stable the oil will be at varying temperatures and conditions.
Extreme Pressure and Wear Protection	ASTM D 3233: Falex Load Test: Measures extreme pressure properties (in pounds) of a fluid lubricant using the Falex pin method. Greater amount equals greater extreme pressure characteristics.
	ASTM D 2783: Film Strength: Uses four-ball method to determine load-carrying properties of lubricating fluid. The higher the number, the better is the protection.
	ASTM D 2783: Weld Point: Measures the lowest applied load in kilograms at which the rotating ball welds to the three stationary balls. The higher the number, the better are the extreme pressure characteristics.
	ASTM D 2783: Load Wear Index: An index of the ability of a lubricant to minimize wear at applied loads. The higher the number, the better is the wear protection.
	ASTM D 2782: Timken Method: Line contact, similar to roller bearing, the higher the number, the greater are the load-carrying capabilities.
	ASTM D 4172: Four-ball (wear scar): Measures the relative wear preventive properties of lubricating fluids in sliding contact, using the four-ball method. The lower the number, the more protection the oil provides.
Corrosion Protection	ASTM D 130: Copper Corrosion: Measures ability to protect yellow metals from corrosion. The range is 1A being the best rating and 4 being worst, 1B is typical.
Foaming Tendency	ASTM D 892: Foam Test: Measures the amount of foam produced at specified temperatures and different air blowing intervals. No foam (0/0) is best.
Oxidation Resistance	ASTM D 189: Conradson Carbon Residue: Measures the residue formed by evaporation and thermal degradation of the oil. The less percentage of carbon residues, the better.

(*Continued*)

Gear Oil Properties, Test Methods, and Standards (Continued)

Gear Oil Property	*Test Methods and Standards*
Low-Temperature Flow and Pumpability	FTM 3456: Channel Point: When a gear passes through the gear oil, the oil flows in behind it filling in the channel left by the gear. This method records the temperature at which the gear oil no longer flows back into the channel.
	ASTM D 97: Pour Point: The lowest temperature at which movement of the fluid is observed is recorded as the pour point. The lower the pour point, the better utility the fluid has for certain applications at low temperature.

Electrical Wire

The Federal Emergency Management Agency (FEMA) reports that electrical distribution was the fifth-ranked cause of fires, the fourth-ranked cause of fire fatalities, and the second-ranked cause of property loss.

Causes of U.S. Residential Fires Due to Electrical Distribution

Cause of Fire	*Percent (%)*
Fixed Wiring	34.7
Cords and Plugs	17.2
Light Fixtures	12.4
Switches, Receptacles, and Outlets	11.4
Lamps and Lightbulbs	8.3
Fuses, Circuit Breakers	5.6
Meters and Meter Boxes	2.2
Transformers	1.0
Unclassified or Unknown Electrical Distribution Equipment	7.3

If you remove installation and selection errors, corrosion is the most common reason for electrical wire failures. Corrosion can be classified based on the appearance of the corroded material. The classifications used vary slightly from reference to reference, but it is generally considered that there are eight different forms of corrosion:

Uniform or General: Corrosion that is distributed more or less uniformly over a surface.

Localized: Corrosion that is confined to a small area. Localized corrosion often occurs due to a concentrated cell. A concentrated cell is an electrolytic cell in which the electromotive force is caused by a concentration of some components in the electrolyte. This difference leads to the formation of distinct anode and cathode regions.

Pitting: Corrosion that is confined to small areas and takes the form of cavities on a surface.

Crevice: Corrosion occurring at locations where easy access to the bulk environment is prevented, such as the mating surfaces of two components.

Filament: Corrosion that occurs under some coatings in the form of randomly distributed threadlike filaments.

Intergranular: Preferential corrosion at or along the grain boundaries of a metal.

Exfoliation: A specific form of corrosion that travels along grain boundaries parallel to the surface of the part causing lifting and flaking at the surface. The corrosion products expand between the uncorroded layers of metal to produce a look that resembles pages of a book. Exfoliation corrosion is associated with sheet, plate, and extruded products and usually initiates at unpainted or unsealed edges or holes of susceptible metals.

Galvanic: Corrosion associated primarily with the electrical coupling of materials with significantly different electrochemical potentials. Saltwater provides an environment for galvanic corrosion to occur, because it helps to establish a galvanic potential due to the electrolytic properties. When dissimilar metals come in contact in the presence of an electrolytic solution, a free flow of electrons occurs. The surface of the metals becomes oxidized due to a transition that takes place. The oxide layer, albeit stable, lacks the physical integrity of the original metal. The oxide layer does not conduct electricity or heat very well. It is very possible that the corrosion facilities overheating and eventual failure.

Environmental Cracking: As discussed in the previous sections, this occurs partially due to the corrosive effect of an environment.

Corrosion Fatigue: Fatigue cracking characterized by uncharacteristically short initiation time and growth rate due to the damage of corrosion or buildup of corrosion products.

High-Temperature Hydrogen Attack: The loss of strength and ductility of steel due to a high-temperature reaction of absorbed hydrogen with carbides. The result of the reaction is decarburization and internal fissuring.

Hydrogen Embrittlement: The loss of ductility of a metal resulting from absorption of hydrogen.

Liquid Metal Cracking: Cracking caused by contact with a liquid metal.

Stress Corrosion: Cracking of a metal due to the combined action of corrosion and a residual or applied tensile stress.

Fretting Corrosion: The protective layer becomes compromised, and damage occurs at the interface of two surfaces. Exposure of the metal coupled with an increase of abrasive particles and oxidation leads to fretting corrosion.

Noncorrosion failure modes include the following:

Fusing Failure: Excessive current can cause copper magnet wire to fuse or melt, and the ends of the failed wire can reveal what happened. Typically, the rate of increase, or rise time, of the current can affect the shape of the wire end. If current increases slowly to the conductor's fusing limit, wire ends generally appear rounded and show signs of burned insulation and slightly melted or bulged copper—the only visual signs of the jacket burning. But, if the current increased suddenly (from a short circuit), a small ball of copper on one of the fused wire ends would be visible.

Tensile Failure: The wire "thins" out at certain sections due to strain on the wire—elongation occurs. A loose lead wire or terminal can transmit excessive tension to the wire. This type of failure usually occurs close to a terminal or lead-wire attachment point.

Metal Fatigue: Caused by repeated bending, produces a rough, flat end when the wire fails. The jacket will have pulled away from the conductor. A tell-tale sign of metal fatigue is that there does not appear to be an oxide layer on the surface of the metal. A nick, scratch, or abrasion in the wire surface concentrates stress at this point and increases the susceptibility of the wire to tensile or bending stress.

Tool Cuts: Nicks on a wire from a tool (cutters, screwdrivers, pliers, etc.) can facilitate failure. Signs of this sort include a wire that is cut transversely, which is normally due to a knife. Wire cutters leave a ridge on the end of the wire. There will be equal angles on each side of the cut end pushing the insulation into a cut. Shear cutters will produce score marks, and the insulation will appear to be compressed.

Electric Wire Physical Properties

Nominal diameter (inch)
Rated voltage (kilovolts)
Rated temperature (°C)
Cable Shielding
Cable Jacket
Cable Insulation

Electric Wire Physical Properties (Continued)

Cable Conductor
Conductor Size
North American (AWG)
> AWG 0
International (mm^2)
Conductor type
International (mm^2)
Number of strands
Flat conductor size
Width (inch)
Thickness (inch)
Conductor weight (lbs/ft)

Electric Wire Properties, Test Methods, and Standards

Properties	*Test Methods and ASTM Standards*
General Specification for Aluminum Wire	B233 Specification for Aluminum 1350 Drawing Stock for Electrical Purposes
General Specification for Thermoplastic Wire Jacket	D2308 Specification for Thermoplastic Polyethylene Jacket for Electrical Wire and Cable
Resistivity	B193 Test Method for Resistivity of Electrical Conductor Materials
Size (Diameter)	B258 Specification for Standard Nominal Diameters and Cross-Sectional Areas of AWG Sizes of Solid Round Wires Used as Electrical Conductors
Tensile Strength	B557 Test Methods for Tension Testing Wrought and Cast Aluminum and Magnesium-Alloy Products
	B557M Test Methods for Tension Testing Wrought and Cast Aluminum and Magnesium-Alloy Products (Metric)

Lighting

There are many causes for a lightbulb to fail. For standard incandescent bulbs, failure results when the filament breaks or the bulb breaks. Long-term failure occurs when the filament eventually becomes so fatigued that its electrical resistance increases to the point that current will not flow. The inside of the bulb gets very dark, and the electrical contact on the base starts to burn.

Filament Breakage (occurs due to):

- Evaporation of the metal
- Too much current, which overloads the capacity of the filament
- Vibration
- Frequently turning the light on and off creates a temperature change that occurs every time that happens, resulting in thermal imbalance and shocking.
- Improper size and construction of the filament

Thermal Shock: The most common failure mode of bulbs, as soon as the switch is turned on the bulb flashes brightly and then fails. The reason most bulbs fail in this manner is thermal shock. When the switch is turned on, full current suddenly hits the filament at the speed of light. This sudden, massive vibration causes the filament to wildly bounce. This mechanical movement causes metal fatigue (just like bending a paper clip until it breaks), resulting in breakage of the filament. That is why lights left on always last a long time, although if the lights are left on, you will have to put up with listening to your parents ask if you have stock in the power company.

Sealing Failure: Occurs when the bulb's atmospheric seal is broken and oxygen enters the bulb. The filament burns up instantly. Such failures occur when bulbs are screwed into sockets too tightly.

Installation Failure: Sometimes when we twist the lightbulb too tightly into the socket, it fails for another reason. The problem occurs because the contact at the center of the bulb base is not always consistent in size or shape. If the brass lead in the center of the socket does not make full contact with the bulb, an arc can occur, and the resulting heat will melt the solder. Over time, the solder deforms and a gap becomes larger between the lead and the contact. When this happens, the electrical contact between the brass tab and the solder is broken, and the lightbulb acts as if it has been turned off when in fact the switch is on and the bulb filament is fine but the base solder is not.

This can be avoided by making sure the brass tab is always at about a 20° angle inside the bottom of the socket. People who twist bulbs in tightly will depress and flatten the tab so it does not spring back when a bulb is replaced. If you find this to be the case, simply remove the source of energy (Lock Out/Tag Out); in other words, unplug the lamp and bend the tab and slowly pull it up so the end of the tab is about one-quarter inch off the base of the socket.

The proper way to install a lightbulb is to always do so with the power on and the light switch on. As soon as the bulb comes on, turn the bulb one eighth of a turn. If you screw the bulb in too tightly, you will once again flatten the brass tab.

Halogen Installation Failure: Touching halogen bulbs creates hot spots that make the glass heat unevenly and cause failure of the bulb. It is a best practice to use rubber gloves, paper towels, or tissue paper. Please keep in mind that these bulbs get incredibly hot. If you touch them right after they have been blown out or turned off, be warned—you will burn your fingers. Been there, done that.

Light-Emitting Diode (LED) Failure Modes

Typically, an LED will slowly fade before failing completely. An abrupt failure is rare but possible. The following are several failure modes for LEDs:

Active Region Degradation: Occurs in gallium arsenide and aluminum gallium arsenide LEDs. When the crystal has an anomaly or defect, heat becomes the dominant source of energy rather than visible light, which promotes nucleation and further growth of dislocations. This is driven by the crystal shape of these semiconductors. Gallium nitride and indium gallium nitride have a different crystal lattice arrangement that exempts them from this failure mode.

Short Circuits: Caused by the formation of whiskers (which are exceedingly difficult to find after the failure due to a complete meltdown). Whiskers are conductive pathways caused by mechanical stresses, high currents, and even a corrosive environment.

Thermal Stress: When the epoxy temperature rises to its glass transition temperature (Tg), it starts rapidly expanding, causing mechanical stresses on the semiconductor and the bonded contact. The material can also shrink excessively at very low temperatures.

Electrostatic Discharge (ESD): A pulse of intense electromagnetic energy can cause immediate failure of the semiconductor within the LED.

Lighting Physical Properties

Application
Fixture type
Fixture

(*Continued*)

Lighting Physical Properties (Continued)

Voltage (volts)
Wattage (watts)
Lighting fixture dimensions
Lamp type
Lamp base
Lamp power (watts)
Rated average life (hours)
Color temperature (K)
Mean lumens (lm)
Light center length (inch)
Diameter (mm)
Maximum overall length (inch)
Base type
Filament type

Electric Motors

Electric motor failures are very dependent on the application and environment. When a motor fails, it can do so for several reasons. The Institute of Electrical and Electronics Engineers (IEEE) investigated the top six reasons for electric motor failure:

1. Bearings: 51%
2. Stator windings (from voltage surge, water, overload): 16%
3. External (environment, voltage, load): 16%
4. Unknown: 10%
5. Rotor bar: 5%
6. Shaft/coupling: 2%

Unfortunately, the actual root causes of these failures are difficult to determine for the same reason why any root cause is difficult to determine—because many involved do not perform proper root cause analysis or implement corrective actions. These statistics are rather general. Therefore, it is vital that a root cause analysis be performed when failure occurs. Motor failure is a result of stresses that acted upon a component and brought about a change in the functionality of the device. By better understanding these stresses, it is more likely the root cause of the failure will be uncovered. When it comes to motors, the root cause can be rather obvious, for instance:

- The winding was struck by debris from a balancing weight that came loose.
- Water completely saturated the winding.
- The grease in the bearing became contaminated.

In most cases, it is not that simple. Motor failure is a very serious matter, on which much work has been done to identify the various types of stresses as well as the components that fail due to the stresses.

- Components
  - Bearings
  - Stator
  - Rotor
  - Shaft
  - Frame
- Electric motor stress
  - Thermal
  - Electric/dielectric
  - Mechanical
  - Dynamic residual
  - Electromagnetic
- Environmental

Most electric motor failures are caused by a combination of various stresses that act upon the bearings, stator, rotor, and shaft. If these stresses are kept within the design capabilities of the system, premature failure should not occur. However, if any combination of the stresses exceeds the design capacity, the life of the system may be drastically reduced, and catastrophic failure could occur.

These stresses are classified as follows:

Bearing Stresses—Thermal, dynamic, and static loading; vibration and shock; environmental; mechanical; and electrical as they relate to the following:
- Motor design
- Materials of construction
- PM/PdM effectiveness
- Insulation class
- Bearing type and quality

- Enclosure type
- Quality of construction
- Operating environment
- Load characteristics

Stator, Rotor, and Shaft—Thermal, dynamic, mechanical, environmental, magnetic, residual, and miscellaneous.

Motor—A number of issues contribute to excess bearing loading:

- Wrong coupling type or installation
- Dynamic overloading (balance, misalignment)
- Belt, sheave, or insert wear
- Belt misalignment
- Incorrect belt tension
- Static loading
- Poor workmanship
- Defective bearing housings

Electrical—Heat in the winding breaks down the insulation, which leads to failure. The IEEE reports that insulation failure contributed to 30% of failures, and, of these, 60% were caused by overheating. It is important to focus on eliminating every potential source of overheating. Other causes of insulation breakdown include attack by dust, moisture, or chemical agents, and contaminated windings. Heat and other causes such as vibration or mechanical shock might induce electrical problems. The power supply also has the potential to cause motor failure.

Electric Motor Physical Properties

DC Motors
DC voltage (VDC)
Shaft speed (rpm)
Continuous current (amps)
Continuous torque (In-lbs)
Output power (HP)
Torque constant (in-oz/amp)
Rotor inertia (oz-in-sec^2)
DC construction
Commutation
Shaft orientation

Electric Motor Physical Properties (Continued)

DC Motors
Gearing Options
Gearing
Gearhead model
Gear type
Gearbox ratio (:1)
Gearing efficiency (%)
Housing/Enclosure
Units
Motor shape
Diameter/width (inch)
Length (inch)
NEMA frame
Options
Extreme environment
Environment
Operating temperature (°F)
Shock rating (g)
Vibration rating (g)
Environment
AC Motors
AC voltage
Phase
Shaft speed (rpm)
Continuous current (amps)
Continuous torque (In-lbs)

(*Continued*)

Electric Motor Physical Properties (Continued)

AC Motors
Output power (HP)
Rotor inertia (oz-in-sec^2)
Motor Type
AC construction
Features
Shaft orientation
Gearing Options
Gearing
Gearhead model
Gear type
Gearbox ratio (:1)
Gearing efficiency (%)
Housing/Enclosure
Units
Motor shape
Diameter/width (inch)
Length (inch)
NEMA frame
Options
Extreme environment
Other Specifications
Feedback
Features
Environment
Operating temperature (°F)
Shock rating (g)

Electric Motor Physical Properties (Continued)

AC Motors
Vibration rating (g)
Environment
Linear Motors
Continuous thrust (lbs)
Peak force (lbs)
Maximum speed (in/sec)
Maximum acceleration (g)
Continuous current (amps)
Current/phase (amps/phase)
Motor force constant (lbs/amp)
Dimensions
Stator length (inch)
Travel (inch)
Width (inch)
Length (inch)
General Specifications
Design units
Stepper resolution (steps per inch)
Number of leads
Maximum coil temperature (°F)
Features
Environment
Operating temperature (°F)
Shock rating (g)
Vibration rating (g)

(*Continued*)

Electric Motor Physical Properties (Continued)

Stepper Motors
DC voltage (VDC)
Shaft speed (rpm)
Current per phase (amps/phase)
Holding torque (In-oz)
Output power (HP)
Rotor inertia (oz-in-sec^2)
Motor Type
Stepper construction
Step angle (degrees)
Number of leads
Shaft orientation
Gearing Options
Gearing
Gearhead model
Gear type
Gearbox ratio (:1)
Gearing efficiency (%)
Housing/Enclosure
Units
Motor shape
Diameter/width (inch)
Length (inch)
NEMA frame
Options

Electric Motor Physical Properties (Continued)

Other Specifications
Feedback
Features
Environment
Operating temperature (°F)
Shock rating (g)
Vibration rating (g)
Environment
Extreme environment
Servo Motors
AC construction
AC voltage
Phase
Shaft speed (rpm)
Continuous current (amps)
Continuous torque (In-lbs)
Output power (HP)
Rotor inertia (oz-in-sec^2)
Motor Type
Features
Shaft orientation
Gearing Options
Gearing
Gearhead model

(*Continued*)

Electric Motor Physical Properties (Continued)

Servo Motors
Gear type
Gearbox ratio (: 1)
Gearing efficiency (%)
Housing/Enclosure
Units
Motor shape
Diameter/width (inch)
Length (inch)
NEMA frame
Options
Extreme environment
Environment
Operating temperature (°F)
Shock rating (g)
Vibration rating (g)
Environment
Gear Motors
Motor Description
Construction
Features
AC voltage
Phase
DC voltage (VDC)
Commutation
Performance Specifications
Shaft speed (rpm)

Electric Motor Physical Properties (Continued)

Gear Motors
Continuous current (amps)
Current per phase (amps/phase)
Continuous torque (In-lbs)
Holding torque (In-oz)
Output power (HP)
Torque constant (in-oz/amp)
Rotor inertia (oz-in-sec^2)
Gearing Options
Gearhead model
Gear type
Gearbox ratio (:1)
Gearing efficiency (%)
Shaft orientation
Housing/Enclosure
Units
Motor shape
Diameter/width (inch)
Length (inch)
NEMA frame
Options
Extreme environment
Environment
Operating temperature (°F)
Shock rating (g)
Vibration rating (g)
Environment

The following article addresses a serious problem—counterfeits—but not counterfeit money or jewels or fine art. This addresses something far more serious—counterfeit MRO items. At first glance, one may scoff at the idea, yet it is a situation that not only requires attention but action as well.

Counterfeits Can Kill

There's more evidence than ever before that fake products and parts are detrimental to those who purchase them. Don't risk lives, reliability, and profits just to save a buck.

We who live and work in societies with long, strong histories of industrial responsibility and regulatory compliance are accustomed to choosing freely among a variety of sources for components and replacement parts. "Original" or "factory" parts come from the company that built the machine and are understood to be the same as those used on the production line. Original equipment manufacturer (OEM) components are expected to be made by the company that supplies the factory, and therefore equivalent to factory parts in every way but the name. Aftermarket parts come from a variety of sources and are trusted according to the brands under which they are built or sold.

We've chosen our parts based on real or perceived quality, value, supplier reputation or support, and suitability for intended purposes. Some prove better than others, but we've rarely been surprised by non-functioning or dangerous parts. Maybe this is because it traditionally has taken a significant investment in manufacturing equipment and the supply chain to establish a brand, make the parts, and bring them to market. That, plus the requirements of certification agencies such as Underwriters Laboratories (UL), have largely kept low-quality and non-functional components out of our plants. But not anymore.

A Growing Pain

Moving beyond Rolex watches, low-quality bolts and brand-name auto parts, counterfeiters are producing fake industrial equipment and components complete with bogus marks, packaging, documentation, and certifications. Also creeping into the supply chain are imitations or knockoffs that wear their own brands (or none) but deliberately mimic the appearance of famous-maker products. The International Chamber of Commerce (ICC, www.icc.cca.org) estimates that trademark counterfeiting accounts for about 6% of world trade. It's worth an estimated $350 billion annually, according to the white paper, "The Threat of Counterfeit Product Approval Marks Warrants Aggressive Detection and Enforcement Action," by a recent alliance between U.S. Occupational Safety and Health Administration (OSHA, www.osha.gov) and the

American Council of Independent Laboratories (ACIL, www.acil.org). "In its mid-year 2005 report, the U.S. Department of Homeland Security's Customs and Border Protection division reported that customs seized more than $64 million worth of counterfeit products in 3,693 seizures. Electrical equipment, much of it intended for the U.S. workplace, alone accounted for more than $6 million," the paper says. "The International Anti-Counterfeiting Coalition (IACC, www.iacc.org) reports the majority of counterfeit products come from Asia, primarily China, and that Eastern Europe also has become a significant source. The manufacture and distribution of counterfeit products has been linked to organized crime. Counterfeit approval marks have been found on electrical products built with substandard materials and exhibiting compromised electrical spacing—both of which pose potential shock and fire hazards to U.S. employees." Jim Pauley, vice president, industrial and government relations for Schneider Electric (www.us.schneider-electric.com) received a phone call from the U.S. customs office in San Francisco a couple of years ago. "They found a person trying to get through from China with a suitcase full of circuit breakers with our Square D logo and UL labels on them," Pauley says. "All of them were counterfeit. This has led to several ongoing litigations, and six settled lawsuits, but we think the overall state of the problem is still much worse than people realize or are willing to admit that it is." Schneider Electric has uncovered counterfeit activity, but it is difficult to estimate how widespread the issue is. "This is what we know, and there are probably more," Pauley says. "Customs usually inspects about 2% of all cargo, and the rest can't all be crystal clean."

The British Valve and Actuator Association's (www.bvaa.org.uk) technical director reports there was very little counterfeiting of mainstream valves just five or 10 years ago, but now there's more anecdotal evidence than ever before about fake products and parts, coming mostly from Asia and specifically China, says Rob Bartlett, director of the organization. "Everyone has a story." Typically, a defective part will be returned to the manufacturer, often through an authorized distributor. "We find out which contractor returned it, find out where he got it, trace down the source and refer it to the Consumer Product Safety Commission," says Bill Snyder, vice president, channel development, Square D. "It gets traced back to one or two factories in China, which get raided and shut down. But they reopen a few days later, a couple of miles away."

Pauley points out that the parts being copied were never manufactured in China. "Don't misunderstand," he says. "People think this is what happens when you manufacture in China, but these products are not made by us in China. They have counterfeit 'Made in USA' labels on them. This is not a 'manufactured in China' problem, it's a different group. These are criminals under U.S. law."

It would be one thing if counterfeiting only compromised patents, copyrights, and sales revenues. But in the case of industrial components, it's also a matter of functionality. "They put on fake Square D and UL labels, and the parts don't work. A counterfeit breaker subjected to a fault current that any breaker would be expected to clear just blows up," says Snyder. "And this is not limited to Schneider or Square D—our competitors are seeing the same thing with products from wiring devices to cords to allegedly explosion-proof boxes and fittings." Users may not realize this until years later, when those fake devices are called on to work and fail instead.

Wild, Wild Web

One contributor is Internet-based businesses that routinely sell millions of dollars worth of equipment and enable the smallest shop to market and deliver items worldwide. These speedy, global transactions can be helpful, but participants reportedly don't conduct as much verification and certification as traditional manufacturer-to-distributor channels.

"We haven't personally identified any counterfeit items, but our customers report seeing many items on eBay for more than 30% off list prices," says David Stock, a system integrator at Innovative Control Inc. (ICI, www.innovativecontrolinc.com), Crystal Lake, Illinois. "If someone else wants to purchase equipment that way it's fine with me, but I think buying in an environment without distributors and traceability is a serious risk." Some purchasers accept counterfeiting and knowingly buy fake devices. "Some people actually choose to purchase counterfeit products, assuming they're paying less for products that are equal in value to the legitimate products they mimic," states the OSHA-ACIL report. "People who deliberately choose to buy counterfeit products aren't victims. Instead, they support the criminally deceptive practices of counterfeiters by creating a built-in market for their goods. If consumers stopped using counterfeit products, counterfeiting wouldn't disappear. However, in many cases, counterfeiting would be less profitable and more risky without these easy sales." Anthony Todarian of the Canadian Standards Association (CSA, www.csa.ca) adds that it and other agencies regularly issue product alerts and recalls when it finds counterfeit products, and that eBay and other online sellers have promised to remove them when they're notified. Pauley adds it's one thing to buy a contractor's leftover 10-pack of circuit breakers on eBay, but sellers that indicated they have large numbers of breakers for purchase raise significant questions about where they're coming from. "When you see higher volumes and a supplier saying it can deliver any amount, you have to ask if this is a legitimate source," says Pauley. "This is why you can't just run to the Internet to buy circuit breakers. You have to know your sources and start further up the food chain. If you wait until a product is delivered, it's probably too late."

We expect parts to fit, function and endure. But counterfeits generally use cheaper and less safe materials, such as flammable plastics, less durable alloys, loose tolerances, and inadequate electroplating. Copiers cut production costs by not respecting quality and safety standards, imitating external appearances with no knowledge or understanding of the technologies needed to produce functional, durable parts. In the case of gearboxes, "We've seen direct knock-offs that look the same from six feet away," says Bill Engle, plant manager, Dodge Gear, Greenville, South Carolina. "You'll save about 40%, but after 150 hours they catch fire." Some of the knockoffs came back from distributors as "warranty returns." Engle doesn't know how the end users get the knockoffs, but until they get back to the factory, everybody is under the impression they're genuine Dodge gearboxes. Poorly-made bearings contribute to excess friction and wear that cause overheating. "We're seeing some counterfeit bearings from China," says Bill Bayliss, business manager-aftermarket at FMC FoodTech (www.fmctechnologies.com/foodtech). "Some bearings are very sophisticated and, as a result, very expensive. But there's a reason they're so expensive. Some customers are finding out the hard way." Some fake bearings are obvious on inspection. Visible clues can include packaging differences, alternate countries of origin or oddball serial number formats or differences in the look of subcomponents. Names and logos might be misspelled. Increasingly however, the more sophisticated fakes are essentially identical under routine inspection or testing. The hidden differences, determinable only by experts, may have serious consequences. These nonvisible differences can include nonhardened races, alloys without adequate corrosion or wear resistance, unserviceable seals or defective lubrication, suggests Shaeffler KG, maker of INA and FAG bearings. In some cases, the fakes can look "more real" than the legitimate product, according to reports from major bearing manufacturers. In most cases, the final determination can only be made by an authorized distributor or the manufacturer. Falsification of bearing identity has reached such proportions that manufacturers, major customers, and testing and certification organizations worldwide are discussing a "Global Bearing Code of Conduct," and the Japanese Bearing Manufacturers Association has published a poster showing a pair of handcuffs whose one side is a large ball bearing. The poster says, "Counterfeit bearings are illegal. They can cause injury or death. Don't produce them. Don't sell them. Don't buy them."

Cannot Judge by Appearance

Close visual inspection of devices and documentation is supposed to help find counterfeits, but several sources say the external appearance of many fakes is so good that they're almost indistinguishable from their genuine counterparts. "You can't identify counterfeit products by physical appearance," says Snyder. "The counterfeits we are seeing today are indistinguishable in outside appearance—I cannot see the difference."

"We work with the government and customs. We give them brochures on what to look for and the next shipment has fixed it. We say to look for grease on the contactor jaws; they put grease on the jaws."

Even newer identification technologies, such as RFID chips and laser etching, can be quickly adopted by counterfeiters. "Many counterfeits look pretty much like the real thing. They may even have duplicate die marks and moldings that serve no purpose. We have seen situations where manufacturers have added holographic labels to their products, and then the counterfeiters copied it nine months later," says Pauley. "Sometimes our engineers have to take apart devices to compare the legit and counterfeit version, and the fakes ones quickly fail if their performance is tested." Still, the battle for positive identification starts with the naked eye. "We work with the OEMs for a way to authenticate a product," says Jack Walsh, director of sales, Videojet (www.videojet.com). "The first way is by quality—a high-quality product and packaging so you can tell by looking. But counterfeiters are clever, and there isn't always packaging, so we do track and trace. At the low level, it's covert marking on the product itself. The high level is using RFID or other markings as a license plate that travels with the product." Every product move in the supply chain is recorded, and authorized distributors must be able to show traceability back to the source.

"An unauthorized supplier won't be able to provide the history," Walsh says. "This is going on in the automotive industry right now. It's called serialization, and it's being used on safety-critical and high-cost items." Registered part IDs can be accessed through the Internet. "If a distributor gets a suspect part, or Customs sees a load of suspect parts or a warranty claim is questionable, they can call the supplier and check it out," he adds. "We're supplying end-to-end turnkey systems for this. "You can't penalize the distributors and shut them down when they say they didn't know the part was counterfeit. Now the manufacturers have a way for them to know."

Know Your Suppliers

Perhaps the best way to avoid counterfeit devices is to buy from manufacturers and distributors that are well known to your own company and its engineers. However, it's also vital to maintain frequent personal contact with authorized suppliers because counterfeiters can set up false representatives and corporations to support their fake products and documentation. "When you're in a rush, you might not check some certifications as close as possible," says Bartlett. "So, besides checking that documents aren't bitmapped images and telephoning to confirm suppliers' claims and identities, buyers also must be responsible for their devices' audit trails, and make sure where, when and who makes these products."

(Reprinted with permission from Jim Montague, Editor at Large, and Paul Studebaker, CMRP, Editor in Chief, "Counterfeits Can Kill," *PlantServices.com* [http://www.plantservices.com/articles/2008/014.html].)

Summary

At this point, you should have a solid understanding of specific failure modes as well as the physical properties and standards for commonly bought items. This will prove essential for the development of a comprehensive purchasing specification. As the failure mode is understood, then the documentation process can begin in earnest. You have now been introduced to various sources of failure. This information is extremely important and valuable when reviewing with the mechanics and technicians as well as vendors who will be pitching solutions. Having a good grasp of failure sources will help establish a very strong foundation for a performance-based purchasing specification.

Tools for the Specification Development Process

The following work sheets can be used to begin the selection and screening process. In Chapter 5, you will be introduced to the writing and selection process of developing the procurement specification. There will be several examples of various cases studies that were carried out to validate and clarify the total cost of failure as well as the total cost of success. The following is a work sheet for developing a case study. The case study will help you gain insight as well as buy-in from management as well as the mechanics and operators.

Case Study Work Sheet

Your case study layout should look like this:

	Data and Work
Problem: What are the specific problems as they relate to equipment and process?	
Proposed Solution: How will the new tool, material, or device solve these problems?	
Study Layout: Measurements are taken before, during, and after the trial of the new product. 1. What are the conditions being monitored (i.e., temperature, pressure, energy in amperes, rpms, throughput, etc.)? 2. How are the measurements going to be taken? 3. What is the frequency of the measurements? 4. What is the duration of each segment?	
Results: Document the results during the process before, during, and even after the changeover.	
Data Interpretation: How did the implementation of the new tool, material, or device affect the equipment and what does that mean in terms of energy costs, downtime reduction, parts replacement reduction, and labor cost reduction?	

Chapter 5

Finalizing the Purchasing Specification and Selecting Vendors

In this chapter, the specification begins to take shape. Data retrieved from various sources are now used to establish the specification requirements. The goal of Chapter 5 is to build the purchasing specification, which includes defining:

- The scope
- Product application requirements
- Terms and conditions
- Price per units, payments, incentives, and penalties
- Contractual terms and conditions
- Requirements of reply
- Evaluation process
- Selection and audits

This chapter includes a section on "Green Purchasing" and a "Purchasing Specification Development Checklist."

Writing the Purchasing Specification

There are few organizations that do a better job at writing specifications than the U.S. military. Many consider the specifications to be overkill, yet consider that the result of failure of a particular material or device could cost lives, if not sovereignty. In doing research for this chapter, I came upon an interesting military specification, A-A-20267.

A-A-20267

JULY 31.1997

COMMERCIAL ITEM DESCRIPTION

MEAT LOAF WITH BROWN GRAVY, SLICED, FROZEN

The U.S. Department of Agriculture has authorized the use of this Commercial Item Description.

1. SCOPE.

1.1 This Commercial Item Description (CID) covers frozen sliced meat loaf with brown gravy, packed in commercially acceptable containers.

2. SALIENT CHARACTERISTICS.

2.1 Processing: The frozen sliced meat loaf with brown gravy shall be prepared in accordance with good commercial practice.

2.2 Ingredients: The frozen sliced meat loaf with brown gravy shall consist of meat, gravy and spices, and safe and suitable sweeteners and thickeners.

2.2.1 Meat: The meat shall conform to the applicable provisions of the Meat and Poultry Inspection Regulations (9 CFR Parts 301 to 350). Non carcass components (e.g., cheek meat, head meat, ox tails, esophagus, hearts, and by-products) shall not be used.

2.2.1.1 Meat loaf mixture: The meat loaf mixture shall contain ground beef and/or ground pork, ketchup, vegetables, bread crumbs, spices, and seasonings.

2.2.2 Gravy and spices: The gravy shall contain beef products (e.g., beef base, rendered beef fat, and beef stock) sweeteners, thickeners, and spices, including salt and garlic. Beneficial comments, recommendations, additions, deletions, clarifications, etc., and any data which may improve this document should be sent to: Head, Food Quality Assurance Staff, Fruit and Vegetable Division, Agricultural Marketing Service, U.S. Department of Agriculture, STOP 0243,1400 Independence Ave., SW, Washington, DC 20250-0243 or telephone (202) 720-9939 or

FAX (202) 690-0102. FSC 8940 DISTRIBUTION STATEMENT A. Approved for public release; distribution is unlimited.

2.3 Finished Product: The finished sliced meat loaf with brown gravy shall show no evidence of excessive heating (materially scorched).

2.3.1 Appearance: The overall appearance of the meat loaf with brown gravy shall not be dry or mushy. There shall be no foreign color to the product.

2.3.2 Odor and flavor: The frozen sliced meat loaf with brown gravy shall have an odor and flavor characteristic of prepared meat loaf with brown gravy. There shall be no foreign odors or flavors such as but not limited to, burnt, scorched, stale, rancid, or moldy.

2.3.3 Texture and consistence: The texture of the meat loaf shall not be rubbery or tough. The sauce shall not be excessively thin, thick, or pasty.

2.3.4 Meat: The finished sliced meat loaf with brown gravy shall be practically free of the following defects: cartilage, coarse connective tissue, tendons, ligaments, or glandular material. The total weight of all defects shall weigh not more than 1 ounce (28.3 g) per 96 ounces (2.72 kg) of finished product.

2.3.5 Foreign material: All ingredients shall be clean, sound, wholesome, and free from evidence of rodent or insect infestation.

2.4 Age requirement: Unless otherwise specified in the solicitation, contract, or purchase order, the food item shall be manufactured not more than 90 days prior to delivery.

2.5 Analytical requirements: Unless otherwise specified, analytical requirements for the frozen sliced meat loaf with brown gravy shall be as follows:

2.5.1 Monosodium glutamate (MSG) content: The monosodium glutamate content of the frozen sliced meat loaf with brown gravy shall not exceed 1.0 percent per 8 ounces (227 g) of finished product.

2.5.2 Fat content: The fat content of the frozen sliced meat loaf with brown gravy shall not exceed 23.0 g per 8 ounces (227 g) of finished product.

2.5.3 Sodium content: The sodium content of the frozen sliced meat loaf with brown gravy shall not exceed 1190 mg per 8 ounces (227 g) of finished product.

2.5.4 Analytical procedure: One pound (454 g) of finished sliced meat loaf with brown gravy shall be selected at random.

2.5.5 Preparation of sample: The sample shall be prepared according to the Official Methods of Analysis of the AOAC International, Method 983.18(b).

2.5.6 Analytical testing: Analyses shall be made in accordance with the following Official Methods of Analysis of the AOAC International Method:

Monosodium glutamate 970.37
Fat 960.39, 976.21, or 985.15
Sodium 969.23

Dry washing and flame emission procedures are to be used.

2.5.7 Test results: The test results shall be reported to the nearest 0.1 percent for monosodium glutamate content, to the nearest 0.1 gram for fat content, and to the nearest milligram for sodium content. Any result not conforming to the finished product requirements shall be cause for rejection the lot.

3. REGULATORY REQUIREMENTS.

3.1 The meat component and the finished product shall originate and be produced, processed, and stored in compliance with all applicable provisions of the Meat and Poultry Inspection Act and the regulations promulgated there under. The delivered frozen sliced meat loaf with brown gravy shall comply with all applicable Federal and State mandatory requirements and regulations relating to the preparation, packaging, labeling, storage, distribution, and sale of the frozen sliced meat loaf with brown gravy within the commercial marketplace. Delivered frozen sliced meat loaf with brown gravy shall comply with all applicable provisions of the Meat and Poultry Inspection Act and regulations promulgated there under.

4. QUALITY ASSURANCE PROVISIONS.

4.1 Product conformance. The frozen sliced meat loaf with brown gravy provided shall meet the salient characteristics of this CID, conform to the producer's own specifications, standards, and quality assurance practices, and be the same frozen sliced meat loaf with brown gravy offered for sale in the commercial market. The Government reserves the right to require proof of such conformance.

4.2 Quality assurance. When required in the solicitation, contract, or purchase order that product quality or acceptability or both be determined, the Processed Products Branch, Fruit and Vegetable Division, Agricultural Marketing Service, U.S. Department of Agriculture, shall be the certifying activity and shall make the determination in accordance with applicable PPB procedures. The frozen sliced meat loaf with brown gravy shall be examined or analyzed or both in accordance with applicable provisions in this CID, solicitation, contract, or purchase order, and, when applicable, the United States Standards for Condition of Food Containers in effect on the date of the solicitation.

5. PACKAGING.

5.1 Preservation, packaging, packing, labeling, and case marking shall be as specified in the solicitation, contract, or purchase order.

6. NOTES.

6.1 Purchaser shall specify:

The container size for the frozen sliced meat loaf with brown gravy.

6.2 Sources of documents.

6.2.1 Source of information for nongovernmental document is as follows:

Copies of the Official Methods of Analysis of the AOAC International may be obtained from: AOAC International, 481 North Frederick Avenue, Suite 500, Gaithersburg, MD 20877.

6.2.2 Sources of information for governmental documents are as follow:

Applicable provisions of the Meat and Poultry Inspection Regulations are contained in 9 CFR Parts 200 to end. The Federal Food, Drug, and Cosmetic Act are contained in 21 CFR Parts 1 to 199. These documents may be purchased from: Superintendent of Documents, New Orders, P.O. Box 371954, Pittsburgh, PA 15250-7954. Credit card (MasterCard or Visa) purchases may be made by calling the Superintendent of Documents on (202) 512-1800. Copies of the United States Standards for Condition of Food Containers are available from: Chairperson, Condition of Container Committee, STOP 0243, 1400 Independence Avenue, SW, Washington, DC 20250-0243.

Civil agencies and other interested parties may obtain copies of this CID from: General Services Administration, Federal Supply Service, Specifications Section, Suite 8100, 470 East L'Enfant Plaza, SW, Washington, DC 20407.

(I am certain that my wife uses that very specification to produce the meat loaf we all "enjoy.")

The sections of the meat loaf specification are sound: scope, characteristics, process, ingredients, finished product, age requirements, analytical requirements, regulatory, quality assurance, and packaging. The only thing missing is the performance requirements. For meat loaf it would be rather difficult to assign performance due to an inability to define failure (short of botulism or ptomaine poisoning). The recipe addresses performance and is objective. I am sure that in some distant galaxy there are life forms that would consider my wife's meat loaf of "high performance"—perhaps as a weapon or energy source when burned. Here on earth, the catfish in our pond turn their whiskers up at it when used as bait. Maybe they are spoiled on pizza.

Following is an example of a government purchasing specification for pipe that has been circulating through many government and military facilities for years.

GOVERNMENT PURCHASING SPECIFICATIONS FOR PIPE

1. All pipes are to be made of a long hole, surrounded by metal or plastic centered around the hole.
2. All pipes are to be hollow throughout the entire length—do not use holes of different length than the pipe.
3. The I.D. (inside diameter) of all pipes must not exceed the O.D. (outside diameter)—otherwise the hole will be on the outside.
4. All pipes are to be supplied with nothing in the hole so that water, steam, or other stuff can be put inside later.
5. All pipes should be supplied without rust—this can be more readily applied at the job site.
   Consider that Some Vendors are now able to supply pre-rusted pipe. If available in your area, this product is recommended, as it will save a lot of time on the job site.
6. All pipe over 500 ft (153 m) in length should have the words "long pipe" clearly painted on each end, so the Contractor will know it is a long pipe.
7. Pipe over 2 miles (3.2 km) in length must have the words "long pipe" painted in the middle, so the Contractor will not have to walk the entire length of the pipe to determine whether or not it is a long pipe.

This example is not real, but we all have had experiences that make us scratch our heads and wonder, "What the heck are they thinking?" Purchasing specifications can be written that will actually reduce overall operating costs. Before we can develop a sound procurement specification, the practice of buying has to be understood.

A specification is a set of specific requirements or parameters required for a tool, instrument, or material to provide an anticipated level of performance. Once the purchasing department as well as maintenance and operations become well versed in developing performance-based procurement specifications, the buying process will become a cost-avoidance process rather than an expense. There are several types of specifications:

- *Performance Specifications*: Describe in detail the product/service in terms of form, fit, and function. The term of form addresses the physical characteristics of the item or service. Fit relates to how the item or service will work in the existing system. Function is defined as what performance levels the item or service is looking to achieve.

- *Design Specifications*: Describes an item or service in terms of a specific material, dimensions, design concepts, methods of manufacture, and so forth.
- *Formulation (Material) Specifications*: This details the chemical composition of the material or item. In addition, it may define the manufacturing or production techniques of the metals, chemicals, plastics, paints, and so forth, that are called out in the specification.
- *Brand Name or Equal Description Specifications*: Identifies products by their brand name, make, and model or catalog number. Often, government bids will use this form, provided applicants can propose an equal product that has the same characteristics as the brand name product or service as is called for in the specification.
- *Specific Make and Model Specifications*: Differ from the Brand Name or Equal Description Specification in that potential vendors cannot propose an equal item. A specific make and model specification restricts competition, must be accompanied by a justification for other than full and open competition, and must be in accordance with corporate regulations and local, state, or federal laws.
- *Hybrid Specifications*: May combine the Performance, Design, and even Formulation Specifications. These specifications are typically the lengthiest to prepare but also offer the highest possible control for overall cost savings and improved reliability.

Specifications should not be confused with standards, descriptions, measurements, or guidelines. Standards are engineering and technical limitations and applications of tools, instruments, machines, or materials. They are the essential criteria needed to establish the status quo of performance for a given item. A standard ensures uniformity and seamless replacement ability. Standards are used in specifications. The American Society for Testing and Materials (ASTM) provides the broadest spectrum of standards, although certain professional organizations have developed comprehensive performance standards as well. These industry-focused standards are typically developed for single-purpose applications. It is not uncommon for an original equipment manufacturer (OEM) to have several talented engineers serve on the technical specification board. The basic outline for a specification is as follows:

- Introduction
  - The Scope of the Requirement
  - The Definition of the Product
  - Functional and Data Standards Cited

- Product Requirements
  - Look and Feel Requirements
  - Usability Requirements
  - Performance Requirements
  - Operational and Environmental Requirements
  - Maintainability and Support Requirements
  - Security Requirements
  - Legal Requirements
- Procurement Requirements
  - Packaging Requirements
  - Shipping Costs and Requirements
  - Product Costs
  - Documentation and Training Requirements
- Reordering Requirements

After a specification has been drafted and several potential suppliers have submitted their replies, the approval process can take place. If a completely new tool, instrument, machine, or material is being considered, then a sample or proof-of-concept (POC) sample should be evaluated. Suppliers wishing to secure future business may provide an expensive tool, device, or instrumentation for trial purposes. This practice has proven to secure the business for years to follow. Many companies test products on a small scale prior to establishing a vendor to ensure a good fit between the company and the vendor. Samples will allow companies to review the product for reliability, efficiency, and effectiveness. The product specifications are often modified during the testing phase. Samples can serve as an excellent way for organizations to completely evaluate the product before making a final vendor decision. An excellent way to select the best supplier is to take the top two vendors from the request for proposal (RFP) process and have them complete the same product. This will demonstrate their project management capabilities, communication style, and ability to meet deadlines for deliverables. Many companies use this technique to shake out the supplier with the most to offer. Perceptive suppliers often catch wind of this technique early in the process. If the supplier judges the potential customer as a growth target, the supplier may allocate extensive resources to inundate and surround the potential customer. Large companies with deep resources have been known to do this, provided they can act nimbly. In many documented case studies, the smaller company willing to provide high performance and quality as well as go beyond by providing unique and valuable value-added contributions will win the account. Many large companies are bogged down in bureaucracy, fiefdoms, and infighting. This provides the opportunity for the smaller company to swoop in and earn the business.

In many cases, companies are now looking to understand the business of the suppliers. A process called due diligence supports or invalidates vendor-supplied information on processes, financials, experience, and performance. Many companies are now using the due diligence process to determine what the potential supplier can do right now, as opposed to what it might do if given the business. This can be an important consideration. The outline of a due diligence includes the following:

- Company profile and history
- Financial status
- Customer references
- Technical, sales, and management structure and background
- Criminal background checks
- Process capability and capacity
- Manufacturing statistics
- Quality initiatives and certifications
- Technology investments
- Security and audit controls
- Legal and regulatory compliance, outstanding complaints or litigation
- Insurance
- Hold harmless agreements

Start-up companies will provide all the necessary paperwork and feasibility reports along with product specifications and even a prototype, but when it comes down to actual experience and capability to consistently produce, they may not be able to come through. It should also be noted that many small companies would work exceedingly hard at securing the business in order to provide an attractive proposition with experience and references. If given the chance, these smaller companies can provide world-class products and services that larger companies cannot or will not. The due diligence process should confirm the information supplied in the specification.

The purchasing department does not want to pour through an endless stack of papers to get to the important aspect, and the prospective vendor does not want to submit the proposal that may hurt chances of winning the business just because the format was difficult to work with. The easiest format that the procurement team can request is in table form with numeric or short answers for the prospective vendor to fill in. This makes the submission process easy, but more importantly, it makes the selection process very easy. The data can be stored in a spreadsheet and quickly analyzed. By following the basic steps of the procurement process, increased cost savings can be realized. It is important that the engineering, maintenance, and even purchasing departments become familiar with failure mode effect analysis (FMEA) and ASTM standards. By pressuring vendors to deliver more in the way of value and performance, increased plant reliability and decreased cost are almost guaranteed.

Often the customer has to use the product and delivery schedule according to the way in which the vendor can package and ship as inexpensively as possible. This may not work for certain companies that require special packaging and delivery schedules and will actually end up costing the company more in the end.

Specification Further Defined

Scope or subject area should be a statement that identifies:

- Who you are.
- What it is you are looking to buy.
- Reasons why you have to buy.
- Terms of the specification.
- How it is to be received.
- How it is to be evaluated.
- When the selection will be finalized.
- Who to contact with questions and responses.

The scope has to:

- *Provide a Clear Definition*: Establish the area you want to address and stay in that area. If it is a specific type of adhesive or motor, do not deviate and require something beyond the intent of the specification.
- *Strike a Balance*: It is important that you are not specific or too broad in the requirements. The FMEA will provide directed insight as to the range of properties required. If you try to cover too much information, you may sacrifice depth. It is better to cover a small subject area in detail than to cover a large subject area poorly. If the scope of your proposed document is very narrow, it might be better to include your information as part of another specification.

Before you begin the specification writing effort, search the Internet for similar specifications. Chances are that the other specification may provide insight that you may have missed. It is also possible that the other specification is less than marginal in style and content and may provide little help.

Product Application Requirements

The packaging and lot size, location of batch information, Material Safety Data Sheets, installation and assembly instructions, certificate of quality assurance, and 24-hour response line are examples of product application requirements. Requirements are categorized in several ways: general, functional, performance, and design. These requirements should identify measurable physical, functional, and quality characteristics that meet the requirements. This may include a detailed list of characteristics, such as sizes, physical dimensions, weights, percent and type of ingredients, types and grades of materials, standard of workmanship, or basic design. The text should be clear, simple language, free of vague terms or those subject to variation in interpretation. The use of abbreviations should be restricted to those in common usage and not subject to possible misunderstanding.

- *Shall, May, Should, or Will?*: Use "shall" to express a requirement binding on the contractor or the purchaser. Use "may" or "should" to express nonmandatory provisions. Use "will" to express future requirements or when certain conditions are met.
- *Definition of Terms*: The necessity for definition of terms can usually be avoided with good specifications. However, in those cases where proper interpretation is necessary or agreement as to definition of terms is a key part of the specifications, such definitions should be included.
- *Measurements*: All terms relating to measurements (gauge, capacity, volume, etc.) should be used in accordance with established precedent and trade practice. Review the document and make every effort to replace words with numbers or use words and numbers in combination (one (1) two (2)). Whenever you use numbers instead of words, communication is truly enhanced. All measuring and testing equipment is designed to provide specific numerical answers.

Figures and Tables

Figures, illustrations, graphs, and so forth, can often describe the item more clearly and accurately than text. They should be used as much as possible.

Group Terms

Designations with established precedent and trade practice include, for example, type, grade, class, and composition. These terms are defined as follows:

- *Type*: Implies design, model, shape, and so forth, of commodities.
- *Grade*: Implies quality of a commodity.
- *Class*: Implies mechanical or other characteristics that are not in quality of grade.
- *Composition*: Implies chemical differences in commodities.

Other terms such as style, color, form, weight, size, and so forth, are also used as group terms.

General Operational Requirements

Simple definitions used to explain and direct the vendor toward fulfilling the buyer's interest in a particular item are general operational requirements. The product objective, the operating environment, constraints, and measures of effectiveness and suitability for the given task will be outlined. The general operational requirements will define the basic need and, at a minimum, answer the questions posed in the following listing:

- Where will the item be used?
- How will the item solve the determined failure mode?
- What are criteria to accomplish the requirement?
- How will the item be used?
- How effective or efficient must the product perform in order to meet the objectives?
- What is the desired duration of use?
- What is the environment the item will be exposed to?

Functional Requirements

Functional requirements explain what has to be achieved and identify necessary performance that must be accomplished.

Performance Requirements

The performance is how well a product does its intended function. Performance requirements should take up the bulk of the writing portion of the specification. It is a direct reflection of the work you put in to relieve failure from the last purchase. The product properties can have a requirement range or be left open for the supplier to provide data.

Table 5.1 presents an example of the specific performance requirements for gear oil. You will notice that the table looks familiar. It was introduced in the previous chapter. An additional field was added for the product data requirements to be included.

Often, product data may not be well understood. The specification may not provide a range, per se, but rather may list the required data. This will require the selection committee to ascertain the required level of performance only after all the responses have been provided.

A paint and coating example is presented in Table 5.2.

Table 5.1 Product Data Requirements to Be Included

Gear Oil Property	*Required Gear Oil Standards*	*Product Data Requirements for Gear Oil*
Gear Oil Operating Viscosity	ASTM D 2983: Low Temperature Brookfield Viscosity Test	88,000 cp
	ASTM D 2270: Viscosity Index	110+
Extreme Pressure and Wear Protection	ASTM D 3233: Falex Load Test	2250+ lb
	ASTM D 2783: Film Strength	23,000+ psi
	ASTM D 2783: Weld Point	400+ kg
	ASTM D 2783: Load Wear Index	55+
	ASTM D 2782: Timken Method	60+ lb
	ASTM D 4172: Four-Ball (Wear Scar)	0.30 mm
Corrosion Protection	ASTM D 130: Copper Corrosion Test Method	1B
Temperature Resistance	ASTM D 2070: Thermal Stability Test	Less than 30%
Foaming Tendency	ASTM D 892: Foam Test	No Foam after 10 minutes
Oxidation Resistance	ASTM D 189: Conradson Carbon Residue	1%
Low-Temperature Flow and Pumpability	FTM 3456: Channel Point	Less than –25°F
	ASTM D 97: Pour Point	Less than –10°F

Design Requirements

Customized products are often required as a "build to," "code to," and "buy to" or "how to execute" product. In this requirement, the buyer is looking for specific criteria that will be determined with exact parameters.

Terms and Conditions

The terms specify the length, start date, and end date of the contract (if applicable), and any options for renewal. Terms and conditions of the specification will vary from company to company. It is highly suggested that the legal department provide input on the requirements they deem appropriate for the specifications.

Table 5.2 Paint and Coating Example

Required Paint Property	*Required Paint Standards*	*Product Data*
Abrasion Resistance	ASTM D5181 Test Method for Abrasion Resistance of Printed Matter by the GA-CAT Comprehensive Abrasion Tester	
Bleed Resistance	ASTM D 279-02 Standard Test Method for Bleeding of Pigments	
Blister Resistance	ASTM D 714-02 Standard Test Method for Evaluating Degree of Blistering of Paints	
Burnish Resistance	ASTM D 6736-08 Standard Test Method for Burnish Resistance	
Checking Resistance	ASTMD 660-93 Standard Test Method for Evaluating Degree of Checking	
Crack Resistance	ASTM D 661-93 Standard Test Method for Evaluation Degree of Cracking, ASTM D 772-86 Standard Test Method for Evaluating Degree of Flaking	
Cure Time	ASTM D3732 Practice for Reporting Cure Times of Ultraviolet-Cured Coatings	
Efflorescence	ASTM D 7072-04 Standard Practice for Evaluating Accelerated Efflorescence	
Gloss Evaluation	ASTM D4449 Test Method for Visual Evaluation of Gloss Differences between Surfaces of Similar Appearance	
Viscosity and Flow	ASTM D4040 Test Method for Viscosity of Printing Inks and Vehicles by the Falling-Rod Viscometer	
Volatile Content	ASTM D5403 Test Methods for Volatile Content of Radiation Curable Materials	

The following is a generic example of terms and conditions of a given item. After examining several dozen examples, it is obvious that they all have been written by the same legal team.

GENERIC PURCHASING TERMS AND CONDITIONS

A note that precedes any request for quote (RFQ) or specification Terms and Conditions section shall be the following:

PARTIES. YOUR COMPANY, its affiliates and subsidiaries, including, but not limited to YOUR OTHER DIVISIONS, SUBSIDIARIES or PARTNERS may make purchases under these terms and conditions and the purchasing business unit is referred to herein as "Buyer." The supplier of Goods or services under this contract of sale is referred to herein as "Seller."

Applicable Terms. The following terms and conditions, together with the applicable Purchase Order and any schedule, exhibit or attachment referenced therein or herein shall constitute the contract of sale between the parties (the "Agreement"). Acceptance of this Agreement by acknowledgment, shipment of all or a portion of the Goods or other performance by Seller shall be unqualified, unconditional and subject to and expressly limited to the terms and conditions of this Agreement. All previous offers by Seller are hereby rejected. Buyer shall not be bound by terms additional to or different from those in this Agreement that may appear in Seller's quotations or bids, acknowledgments, invoices or in any other communications from Seller unless such terms are expressly agreed to in a separate writing signed by Buyer. Any acceptance of any portion of this Agreement shall be deemed an acceptance by Seller of all of these terms as written, without alteration.

Price and Payment Terms. Seller represents that the price charged to Buyer for Goods is at least as low as the price charged by Seller to other buyers of a class similar to Buyer under conditions similar to those specified in the Purchase Order and that such prices comply with applicable government laws and regulations in effect at time of quotation, sale or delivery. Seller agrees that any price reduction regarding any Goods that is implemented prior to shipment or rendering of such Goods, will be applied to all Purchase Orders for shipments of Goods following such price reduction. Unless otherwise specified thereon, prices quoted on any Purchase Order or Purchase Order Revision include any and all charges for the Goods ordered (including, but not limited to, any charges for boxing, packing, crating, cartage, taxes or other added charges). Invoices shall be paid in accordance with the terms stated in the Agreement and due dates for payment of invoices shall be computed from the date of receipt of both the Goods and invoices by Buyer.

Delivery, Title Risk or Loss. Seller shall deliver Goods which are in accordance with the specifications provided by Buyer and Seller is not authorized to unilaterally deviate from Buyer's specifications without written approval from Buyer. Time is of the essence for each Purchase Order and deliveries shall be made both in quantities and at times specified by Buyer; failure to do so shall constitute a breach of this Agreement. Seller shall deliver all Goods free and clear of all liens and encumbrances. If requested by Buyer, Seller agrees to furnish to Buyer, as a condition precedent to final payment, a complete release of all liens, together with a certificate by Seller that the release contains the signatures of all those who performed services or furnished materials under this Agreement. With each delivery Seller shall submit a packing list in duplicate. Title and risk of loss and damage to material purchased by Buyer under this Agreement shall vest in Buyer when the material has been delivered to the point specified in Buyer's Purchase Order, unless Buyer and Seller execute a consignment agreement pursuant to which Seller shall maintain title to the Goods following delivery to Buyer's facility until removed from consignment by Buyer, at which time, Buyer shall assume title and risk of loss. Further, title to Goods purchased by buyer under this Agreement may immediately vest in Buyer at any point where Buyer tenders to Seller both (1) payment for the Goods and (2) written notice of Buyer's desire to take title to the Goods. If this Agreement or a Purchase Order issued pursuant to this Agreement calls for additional services including, but not limited to, unloading, installation, or testing, to be performed after delivery, Seller shall retain title (unless Buyer has paid the invoice) and risk of loss and damage to the material until the additional services have been performed. Notwithstanding the foregoing sentence, if Seller is expressly authorized in writing to invoice Buyer for material upon shipment or prior to the performance of additional services, title to such material shall vest in Buyer upon payment of the invoice, but risk of loss and damage shall not pass to Buyer until completion of the additional services by Seller.

Packing and Traceability. All correspondence must include the Purchase Order number, Release/Line number and Vendor Identification number. All shipments must include packing slips indicating contents, part number or description, Purchase Order number, Release/Line number and Vendor Identification number. When multiple packages comprise a single shipment, the package containing the packing slip must be marked, "Packing Slip Inside." Any transportation charges paid by Seller, to which Seller is entitled to reimbursement, shall be added to Seller's invoice as a separate item and the receipted freight bill shall be attached thereto. All returnable containers for which Buyer is to be invoiced shall be invoiced by Seller and full credit rendered to Buyer upon return, at Seller's expense. If Goods are shipped in returnable containers, Buyer shall take title only to the usable portion of such Goods and Seller shall retain title to any residue remaining in such containers. Buyer shall have no obligation to clean or otherwise restore returnable containers. Seller warrants that Seller's system of production and packaging shall be such as will permit traceability of each lot of Goods, and shall include bar coding if so requested by Buyer. Seller warrants that the packaging of the Goods ordered herein shall be in compliance with all laws relating to packaging of such Goods and shall be adequate for the transit of the Goods undamaged so long as the integrity of the container is maintained.

Premium Shipments. If, for any reason, Seller is unable to meet Buyer's delivery requirements Seller shall immediately notify Buyer of its expected duration of the delay and the reasons for such delay. Neither such notification nor an acknowledgment by Buyer shall constitute a waiver of the applicable delivery schedule or any of Buyer's rights under this Agreement. If Buyer requires a more expeditious method of transportation for the Goods other than the transportation method originally specified by Buyer because of Seller's failure or inability to meet the specified delivery schedule, Seller shall, at Buyer's option, (a) promptly reimburse Buyer the difference in cost which may be incurred by Buyer between the more expeditious method and the original method, (b) allow Buyer to reduce its payment of Seller's invoices by such difference, or (c) ship the Goods as expeditiously as possible at Seller's expense and invoice Buyer for the amount which Buyer would have paid for normal shipment. Seller shall also be liable for any direct and/or consequential damages incurred by Buyer resulting from any delay caused by Seller. Premium freight must be so noted on shipping documents.

Default, Cancellation. Buyer may, in Buyer's sole discretion, elect to cancel this Agreement or any part thereof at no cost to Buyer in the event of Seller's Default as hereinafter described. Sellers Default shall include, without limitation: (a) Seller's failure to comply with the specifications, delivery requirements or terms and conditions of this Agreement; (b) Seller's failure to deliver Goods ordered herein in accordance with the delivery and timing requirement or in accordance with Buyer's specifications; or (c) Seller's threatened or actual refusal to deliver Goods for any reason ("Seller's Default"). In the event of Seller's Default under (b) above, the parties acknowledge and agree that such default by Seller will cause Buyer irreparable harm and Buyer shall be entitled to any and all legal and equitable rights and remedies available to it against Seller to remedy such default, including, without limitation, injunctive relief prohibiting Seller from refusing to deliver the Goods. If Seller has actually refused to deliver Goods as set forth in (b) above, the parties stipulate that it will be difficult to ascertain the amount of damages resulting from such default and that Seller will pay to Buyer $ (amount to be determined) per day as liquidated damages for each day that Seller refuses to deliver the Goods. The parties agree that this sum represents a reasonable estimate of damages and does not constitute a penalty. In case of ambiguity in the specifications, drawings or other requirements of the Agreement, before proceeding, it is Seller's obligation to seek clarification from Buyer, whose written interpretation shall be final. Buyer's right to cancel hereunder shall be in addition to all other rights and remedies available to Buyer under this Agreement or otherwise and Buyer shall have no obligation for payment to Seller for work in progress or otherwise incomplete Goods.

Termination of Convenience. Buyer, in addition to all other rights and remedies it may have under this Agreement or otherwise, shall have the right to terminate this Agreement or any Purchase Order, in whole or in part, without cause, upon notice in writing to Seller. Seller shall thereupon, as directed, cease work and deliver to Buyer all completed and partially completed Goods or materials and work in process, and Buyer shall pay Seller the following, which in no event shall exceed the total price provided for herein: (a) the applicable price provided in the Purchase Order for all Goods which have been completed prior to termination and which are accepted by Buyer, or (b) to the extent commercially reasonable, the actual expenditures on the uncompleted portion of the Purchase Order, including cancellation charges paid by the Seller on account of commercially reasonable commitments made under the terminated Purchase Order. Seller warrants that it will take all steps reasonably calculated to mitigate and minimize the cost to Buyer of such termination.

Proprietary Rights; Infringement. Seller undertakes and agrees to indemnify, hold harmless and, if requested by Buyer, defend, at Seller's own expense all suits, actions or proceedings brought against Buyer, its affiliates and subsidiaries or any of Buyer's directors, officers, employees, agents, dealers, customers, or the users of any of the Goods purchased under this Agreement for actual or alleged infringement of any intellectual property right including, but not limited to, copyright, trademark, trade secret, United States or foreign letters patent or other proprietary rights of any third party on account of the use or sale of any such Good alone or in combination with other Goods or materials and Seller expressly waives any claim against Buyer that such infringement arose out of compliance with Buyer's or its customers' specifications and Seller further agrees to pay and discharge any and all judgments or decrees which may be rendered in any such suit, action or proceeding against any indemnified party.

Warranty. Seller warrants to Buyer and Buyer's customers that Goods furnished pursuant to this Agreement will be new, merchantable, free from defects in design (unless manufactured to a design furnished through Buyer). Material, warning requirements, and workmanship will conform to and perform in accordance with the Buyer's specifications for such Goods and all other agreed upon specifications, drawings and samples. Seller further agrees that it shall be solely liable for all claims of a defect (or alleged defect) in material, merchantability, workmanship, warning requirements and design (unless manufactured to a design furnished through Buyer) of the Good, and from failure to meet any such specifications. These warranties extend to the future performance of the Goods. Seller also warrants to Buyer and its customers that services will be performed in a first class, workmanlike manner. In addition, if Goods furnished contain one or more Sellers' warranties, Seller hereby assigns such warranties to Buyer and its customers. All warranties shall survive inspection, acceptance and payment and shall continue, at a minimum, for the longer of thirty-six (36) months or such period as Buyer has warranted such Goods, or other items of which the Goods are a component, to its customer. Goods or services not meeting the warranties will be, at Buyer's option and without limitation of Buyer's other rights and remedies under this Agreement or otherwise, returned for or subject to refund, repaired, replaced or re-performed by Seller at no cost to Buyer or its customers and with transportation costs and risk of loss and damage in transit borne by Seller. Repaired and replacement Goods shall be warranted as set forth above in this clause.

Quality Assurance. If Seller supplies Goods for use in production under ISO9000, QS9000, AS9000, VDA 6.1, TS16949 or any other quality assurance system specified by Buyer or its customers, Seller shall comply with such quality system standard for such Goods covered by any Purchase Order. Seller agrees to permit Buyer or its customers to review Seller's procedures, practices, processes and related documents to determine such acceptability. This requirement is in addition to any special quality assurance provisions which may be incorporated elsewhere in this Agreement. Records of all inspection work by Seller shall be kept complete and available to Buyer or its customers during the term of this Agreement and for such longer period and in such manner as may be specified by Buyer or required by law.

Inspection, Rejection of Goods. All Goods furnished hereunder and all records to be furnished therewith shall be subject to inspection at destination, notwithstanding any previous inspection, and Seller shall be given notice of any defects other than latent defects within a reasonable time after receipt of the Goods. Buyer may reject or require the prompt correction, in place or otherwise, of any Goods which are defective in material, workmanship, design (unless manufactured to a design furnished through Buyer) or which otherwise fail to meet the requirements of the applicable Purchase Order. Buyer may, in addition to any rights it may have by law, prepare for return shipment and return the Goods to Seller or require Seller to remove them, and the expense of any such action, including, transportation both ways, if any, shall be borne by Seller. If Seller fails promptly to remove such Goods or to proceed promptly to replace or correct them, Buyer may replace or correct such Goods at the expense of Seller, including any excess cost. Payment for any or all of the Goods or services supplied hereunder shall not constitute acceptance by Buyer. Nothing in this paragraph shall in any way limit Buyer's rights under the paragraphs hereof entitled "Warranty" or "Indemnification."

Indemnification. Seller hereby agrees to indemnify, hold harmless and, if requested by Buyer, to defend Buyer and its affiliates and subsidiaries or any of Buyer's directors, officers, employees, agents, dealers, customers, or the users of any of the Goods purchased under the Agreement against and from any and all claims, costs, losses, liabilities and damages (including expenses relating to defense, such as attorney's fees and expenses) arising from a defect or an alleged defect (including, without limitation, failure to warn) in the Goods or other breach of this Agreement, whether such liability arises as a matter of contract (e.g., warranty, repair, replacement, downtime of a customer's assembly line, recall, etc.) or tort (injury to property or person), including, without limitation, all liability for incidental, consequential or special damages. Buyer may, at its option, tender the defense of any claim of liability against Buyer to Seller, in which case Seller shall have the right to settle any such claim provided such settlement is at Seller's expense and involves no action or forbearance by Buyer. Buyer retains the right to defend such claim itself, but subject to indemnification by Seller. Buyer and Seller agree to cooperate reasonably in any such defense.

Right of Inspection. Buyer shall have the right from time to time to send to Seller's manufacturing facilities its personnel for performing tests upon the material or Goods covered by any Purchase Order to ascertain that specified quality standards are being maintained. Buyer's personnel shall have the privilege of visiting all places within the various facilities where raw materials, components or equipment are stored or where manufacturing is being accomplished incident to fulfilling any Purchase Order. Buyer's personnel shall also have the privilege of using Seller's test equipment for the purpose of performing necessary tests.

Supplementary Information. Any specifications, drawings, notes, instructions, engineering notices, technical data, or terms and conditions of Buyer's customer referred to in the Agreement shall be deemed to be incorporated herein by reference as if fully set forth. In case of any discrepancies or questions, refer to Buyer's purchasing department for decision or instruction or for interpretation.

Proprietary Property of Buyer. All specifications, blueprints, technical documents, instructions, molds, models, casts, formulas, sketches, drawings, manufacturing processes, know-how, software and software protocols, electronic commerce system information, inventory management system information, and other business information supplied to Seller under this Agreement or prepared for Buyer under this Agreement shall be proprietary to Buyer ("Buyer's Proprietary Property") and shall remain the sole property of Buyer, except that exclusive designs developed by Seller prior to the placement of a Purchase Order shall remain the property of Seller. Buyer's Proprietary Property shall be confidential, shall not be used by Seller, its agents, representatives or employees for any purpose except in connection with the work to be done by Seller for Buyer under this Agreement, and shall not be used disclosed or made available to any other third party by Seller or its agents, representative or employees. By its acceptance of this Agreement, Seller agrees to take all necessary precautions against theft, destruction, damage, loss, unauthorized duplication or wrongful distribution, or unauthorized use of Buyer's Proprietary Property. Unless otherwise agreed to by

Buyer in writing Buyer's Proprietary Property shall be returned to Buyer upon completion of production or processing or earlier, upon Buyer's demand.

Information Disclosed to Buyer. Unless specifically provided in this Agreement or expressly agreed to in writing by Buyer, no information or knowledge heretofore or hereafter disclosed to Buyer, in the performance of or in connection with this Agreement, shall be deemed to be confidential or proprietary, and any such information or knowledge shall be free from any restrictions (other than a claim for patent infringement) as part of the consideration for this Purchase Order.

Waiver. The failure of either party at any time to enforce any right or remedy available to it under this Agreement or otherwise with respect to any breach or failure by the other party shall not be construed to be a waiver of such right or remedy with respect to any other breach or failure by the other party.

Property Furnished by Buyer. Except as specified below, all patterns, dies, molds, tools, models, jigs, core boxes, piece parts, samples, materials, drawings, specifications, test reports, technical material, advertising material, and any other personal property furnished to Seller by Buyer, or specifically paid for by Buyer for use in performance of a Purchase Order, shall be and remain the property of Buyer, shall be subject to disposition according to Buyer's Instruction and shall be used only in filling orders from Buyer. Any such property furnished by Buyer to Seller shall be appropriately maintained by Seller in order to preserve the condition of such property to the greatest extent possible, reasonable wear and tear excepted. Any waste materials or byproducts generated by or resulting from operations on, use of, or processing of materials furnished to Seller by Buyer, or materials specifically paid for by Buyer for use in performance of a Purchase Order, shall be and remain the property of Seller and shall not be subject to disposition to Buyer's instruction, unless Buyer has specifically requested, in writing, return of such waste materials or byproducts, which in such case will be the property of Buyer.

Insurance. Seller agrees that it will maintain primary, worldwide (when appropriate) insurance in an amount not less than $1 million per occurrence, combined single limit for death, bodily injury and property damage against all liability arising out of the manufacture, sale and use of Goods sold by Buyer, regardless of the date of the occurrence creating such liability. Buyer shall be named as an additional insured under a broad form vendor's endorsement to such policy. Seller will provide Buyer with a certificate of such insurance. At least seven (7) days prior to the start of work on Buyer's premises, Seller shall submit copies of certificates of insurance and policies from insurance companies acceptable to Buyer, for the following types of coverage and minimum limits: (1) Worker's Compensation and Occupational Disease Insurance, and U.S. Longshoremen & Harbor Workers' Compensation Insurance (where required), in statutory limits in accordance with applicable local and federal laws. (2) Employer's Liability Insurance in a minimum limit per person of $1,000,000. (3) Automobile Liability Insurance with a minimum Combined Single Limit of $ 1,000,000 covering all owned, non-owned and hired vehicles used by Seller in the performance of services hereunder. (4) Commercial General Liability Insurance for Bodily Injury and Property Damage covering premises, operations, contractual liability, products, completed operations, and personal injury (false arrest, false imprisonment, malicious prosecution, defamation of character, libel, or slander) in minimum limits of $ 1,000,000 General Aggregate, $1,000,000 Products Completed/Operations

Aggregate, and $1,000,000 Each Occurrence. All certificates of policies furnished must include provisions that no material change or cancellation of the policy will be made without thirty (30) days prior written notice to Buyer. It is Seller's responsibility to determine the adequacy of any subcontractors' insurance and indemnification obligations

Work on Buyer's Premises. In addition to other terms contained herein, if this Purchase Order requires Seller to perform any services upon property (real or personal) owned or controlled by Buyer, the following shall apply: (a) Seller agrees to furnish to Buyer, as a condition precedent to final payment, a complete release of all liens, together with a certificate by Seller that the release contains the signatures of all those who performed services or furnished materials under this Purchase Order. (b) Seller agrees to indemnify, defend and hold harmless Buyer, and its directors, officers, employees and agents, from and against any and all claims and demands (including costs, litigation expenses and counsel fees incurred in connection therewith) arising out of injury to, or death of, any person whatsoever or injury or damage to property of any kind by whomsoever owned, or the environment, arising out of the performance by Seller, Seller's subcontractors or Seller's agents of any work which is the subject of the Purchase Order.

U.S. FASTENER QUALITY ACT; TREAD ACT. In the event that the U.S. Fastener Quality Act (FQA) or the Transportation Recall Enhancement, Accountability, and Documentation (TREAD ACT) applies to any Goods furnished under this Agreement, Seller shall comply with all requirements of the FQA and TREAD ACT and applicable regulations, including without limitation, regulations pertaining to manufacturer's insignia, manufacturer's record of conformance, and record keeping. Seller represents and agrees that all fasteners furnished under this order which are covered by the FQA will have been manufactured in accordance with the FQA. Seller agrees to furnish to Buyer (or Buyer's customers if requested by Buyer) a manufacturer's record of conformance as necessary in support of compliance with the FQA and the TREAD ACT. Seller agrees that any such record (or copies thereof) may be furnished by Buyer to its customers or other parties requiring such documents.

Compliance with Laws. Seller represents that the Goods covered by this Agreement, together with their containers and other packaging, have been manufactured in accordance with the requirements of all applicable federal, state, local and foreign laws, ordinances, regulations and codes ("laws and regulations") and safety constraints on restricted, toxic and hazardous materials, as well as environmental, electrical, and electromagnetic considerations applicable to the country of manufacture and sale. Seller further agrees to furnish Buyer, upon request, a certificate attesting to such compliance in such form as Buyer may require. Seller and all persons furnished by Seller shall comply at their own expense with all such applicable laws and regulations from which liability may accrue to Buyer for any violation thereof by Seller, and including the identification and procurement of required permits, certificates, licenses, insurance, approvals and inspections in performance under this Agreement. Seller agrees to indemnify, defend (at Buyer's request) and save harmless Buyer, its affiliates, its and their customers and each of their officers, directors and employees from and against any losses, damages, claims, demands, suits, liabilities, fines, penalties and expenses (including reasonable attorney's fees) that arise out of or result from any failure to do so.

NAFTA; CERTIFICATION OF ORIGIN; DUTY DRAWBACK. With respect to all Goods delivered from any point within the NAFTA territory (Canada, Mexico and the United States of America), Seller shall provide, with its invoice, a North American Free Trade Agreement Certificate of Origin on U. S. Customs Form 434 or the corresponding Canadian or Mexican form. Seller agrees to transfer to Buyer all customs duty and import drawback rights, if any (including rights developed by substitution and rights which may be acquired from Seller's suppliers), related to the Goods and which Seller can transfer to Buyer. Seller agrees to inform Buyer promptly of any such rights and to supply all documents which Buyer may request or which may be required to enable Buyer to obtain such customs duty and import drawback rights. Seller shall indemnify and hold harmless Buyer, its subsidiaries and affiliates, its and their customers and each of their officers, directors and employees from and against any losses, damages, claims, demands, suits, liabilities, fines, penalties and expenses (including reasonable attorney's fees) that arise out of Seller's non-compliance with U.S. or foreign customs laws or regulations.

Equal Opportunity. This Agreement shall be deemed to include, to the extent applicable hereto: (a) the Equal Employment Opportunity Clause referred to in Executive Order 11246, as amended, (b) all provisions of 41 CFR 60-250, as amended, pertaining to Affirmative Action for Disabled Veterans and Veterans of the Vietnam Era where the value of Goods or services furnished hereunder exceeds $10,000, (c) all provisions of 41 CFR 60-741, as amended, pertaining to Affirmative Action for Handicapped Workers where the value of the Goods and services furnished hereunder exceeds $2,500, and (d) similar applicable requirements of any state or local law.

Changes. Buyer may at any time, by written order, make changes or additions within the general scope of this Agreement. If any such change causes any increase or decrease in the cost of, or the time required for, performance of this Agreement, Seller shall notify Buyer in writing, and an appropriate equitable adjustment will be made in the price or time of performance, or both, by written modification of this Agreement. Any claims by Seller for upward adjustment of price or time requirements must be asserted within (30) days after Seller's receipt of notice of the change from Buyer. Nothing herein shall excuse Seller from proceeding with the Agreement as changed.

Publicity, Promoting or Advertising. Seller shall not, without Buyer's prior written consent, issue any news release, advertisement, publicity or promotional material regarding this Agreement, including denial or confirmation thereof.

Insolvency. If Seller ceases to conduct its operations in the normal course of business, including inability to meet its obligations as they mature, or if any proceeding under the bankruptcy or insolvency laws is brought against Seller or commenced by Seller on its own behalf, or if a receiver for Seller is appointed or applied for, or if an assignment for the benefit of creditors is made by Seller, Buyer may terminate this Agreement without liability, except for deliveries previously made or for Goods covered by this Agreement then completed and subsequently delivered in accordance with the terms of this Agreement.

Drafts. Drafts against Buyer will not be honored and C.O.D. shipments will not be accepted unless expressly agreed to in writing by Buyer.

Governing Law. The contract resulting from this Agreement is to be construed according to the laws of the state of the United States from which this Agreement issues, as shown by the address of Buyer printed on the face of this Agreement.

The parties agree that any controversy arising under this Agreement shall be determined by the courts of said state, and Seller hereby submits and consents to the jurisdiction of said courts.

The elements of the terms and conditions section of the specification should always be reviewed with the legal department as to remain compliant with the various statutes and laws that govern your place of business.

Price per Units, Payments, Incentives, and Penalties

State the terms of payment and include duration, also include a statement on any penalties for inadequate performance resulting in related downtime, parts replacement, labor costs, injury, excessive consumption, or lack of compliance.

Contractual Terms and Conditions

List the standard contracting forms and any certifications or product performance and quality assurances required.

Requirements of the Reply

To properly and quickly evaluate information, it is important that you require a specific form or style in which you are to receive information. One of the easiest methods is in the form of a spreadsheet. It is even better if the information can be e-mailed or filled out online. The various property and standards table provided can act as a template for the supplier to use to reply to the request.

Evaluation Process

The evaluation process indicates the procedures and criteria used for evaluating proposals:

- Developing a timeline for the steps leading to the final decision.
- Setting dates for submitting the letter of intent.
- Sending questions.
- Attending the preproposal conference (if volumes warrant it).
- Submitting the offering (see Appendix for form letter examples).
- Explaining the evaluation process (i.e., Leon–Maxwell equation).

Contact Points

Provide a list of people to contact for information or with any other questions. Incorporate their name, title, responsibilities, e-mail, phone number, and address.

Evaluation

Respondents to the specification should be informed that their replies will be evaluated by matrix decision making. There are several ways you can analyze the specification response. The simplest is combining all the information onto a spreadsheet and seeing which vendor provides the best solution. This can become difficult if the product or service you are trying to choose has several indices to rate. Two tools proven useful in the evaluation process would be the use of a decision matrix and the use of the Leon–Maxwell equation for performance standards. It is important to choose a method and describe the method in the specification.

Specification Example

The following is a very elementary example of a purchasing specification. It is also considered a bid for a request for quote (RFQ). Anyway you slice it, is still a document that is designed to solicit a reply from a supplier to meet a requirement put forth by the originator.

DEPARTMENT OF SOLID WASTE

Due by: **May 16, 2006**
ADMINISTRATIVE INFORMATION

1.0 **Introduction and Background**
Outagamie County Department of Solid Waste is seeking proposals for HDPE piping and fittings to be used for a compressed air supply system at the East Landfill site.

Specifications

Material Requirement (Piping and Fittings)
Meet the requirement of API-15LE and Plastic Pipe Institute Designation PE3608. Cell classification of 345464C in accordance to ASTM D3350 and limits and test methods shall be as follows:

Property	Limits	Test
a. Density	0.955	ASTM D1505
b. Melt Index	0.08	ASTM D1238
c. Flexural Modulus (PSI)	>110,000	ASTM D790
d. Tensile Strength at Yield (PSI)	3,200	ASTM D638
e. Hydrostatic Design Basis	1600 psi at 23°C	ASTM D2837

See attached price sheet at the end of the document for size and quantities.

Qualified vendor must be able to supply all items requested. Contract will be made to only one vendor.

A 24-hour advanced notice is required on all deliveries.

Delivery of materials shall be within 30 days or sooner from the date of the order.

Outagamie County must be notified if any item is backordered. Then vendor must contact Outagamie County with the expected ship date of any backorder.

Delivery address will be:
Outagamie County Solid Waste Department
1419 Holland Rd
Appleton, WI 54911

Contact Information

Technical Information

Kurt Pernsteiner
Outagamie County Dept of Solid Waste
1419 Holland Rd
Appleton, WI 54911
920-832-5004
Monday–Friday, 7:30 A.M. to 3:30 P.M.

Purchasing Policy Information
Michael Purchatzke
Outagamie County Purchasing
410 South Walnut Street
Appleton, WI 54911
(920) 832-1683 Monday–Friday, 7:30 A.M. to 3:30 P.M.
purchamc@co.outagamie.wi.us

5.0 **Clarification and/or Revisions to the Specifications and Requirements**

Proposer must examine the RFP documents carefully and before submitting a Proposal may request from the County's contact person(s) additional information or clarification by the date specified in the RFP timetable. A Proposer's failure to request additional information or clarification shall preclude the Proposer from subsequently claiming any ambiguity, inconsistency, or error.

Requests for additional information or clarifications must be made in writing no later than the date specified in the RFP timetable. The request must contain the Proposer's name, address, phone number, facsimile number; RFP title and the name of contact person(s). Fax to Michael Purchatzke at 920-832-1534 or e-mail at purchamc@co.outagamie.wi.us.

The County will issue responses to inquiries and any other corrections or amendments it deems necessary in written addendum prior to the Proposal due date. Proposers should rely only on the representations, statements or explanations that are contained in this RFP and the written addendum to this RFP. Where there appears to be a conflict between the RFP and any addendum issued, the last addendum issued will prevail.

It is the Proposer's responsibility to assure receipt of all addenda. Outagamie County will send addenda by mail to only those Proposer(s) recorded by the

County as having been sent and/or received a copy of the RFP documents from the County. In addition, Proposer(s) may inspect the RFP documents at the place where they are made available. Upon such mailing or posting, such addenda shall become part of the RFP and binding on Proposer(s).

6.0 **County Reservation**

Outagamie County openly solicits the best possible value on all of our "Requests for Proposals." Because we are a local government, we are able to purchase directly from many of the state and federal contracts. However, in order to not discriminate against our local proposers, we openly solicit proposals of similar pricing structure from all qualified proposers. In the event that all proposals received are in excess of any existing state or federal contract that is available to Outagamie County, we may at our discretion, reject all proposals, and purchase directly from the vendor awarded the state or federal government contract. Outagamie County reserves the right to accept or reject, any or all proposals, in whole or in part, as deemed in the best interest of the County.

a. This proposal request does not commit Outagamie County to make an award or to pay any costs incurred in the preparation of a proposal in response to this request.

b. The proposals will become part of Outagamie County's files without any obligation on Outagamie County's part.

c. The proposer shall not offer any gratuities, favors, or anything of monetary value to any official or employee of Outagamie County for any purpose.

d. The vendor shall report to Outagamie County any manufacturer product price reductions, model changes, and product substitutions. No substitutions are allowed without prior approval from Outagamie County.

e. Outagamie County has the sole discretion and reserves the right to cancel this proposal and to reject any and all proposals received prior to award, to waive any or all informalities and or irregularities, or to re-advertise with either an identical or revised specification.

f. Outagamie County reserves the right to request clarifications for any proposal.

g. Outagamie County reserves the right to select elements from different individual proposals and combine and consolidate them in any way deemed to be in the best interest of Outagamie County.

7.0 **Closing Date**

The County of Outagamie, Wisconsin, will receive proposals up to 2:00 P.M., local time, May 16, 2007. Time may be determined by the U.S. Official Time Clock from the internet http://www.time.gov/timezone.cgi?Central/d/-6/java. Proposals must be delivered to:

Outagamie County Purchasing
4th Floor County Administration Building
410 South Walnut Street
Appleton, WI 54911

The envelope containing your proposal shall show the name of the proposer and must be clearly marked in the lower left hand corner "Proposal—'HDPE Piping and Fittings.'" Any proposal or unsolicited amendments to a proposal received after the closing date and time will not be considered.

8.0 **Facsimile or E-mail of Proposals**
Facsimile will be accepted at 920-832-1534. E-mail will be accepted at purchamc@co.otuagamie.wi.us.

9.0 **Taxes**
Outagamie County is exempt from Federal Excise Tax (39-6005724), Wisconsin Sales Tax (ES 41005), but if there is a tax, such as local or county, it must be shown in the proposal.

10.0 **Venue**
This agreement will be governed and construed exclusively according to the laws of the State of Wisconsin. This agreement is performable in Outagamie County.

11.0 **Status of Proposal**
Proposal results will be posted on Outagamie County's Web site www.co.outagamie.wi.us go to Status of Bids/Proposals.

PROPOSAL FORM

HDPE Piping and Fittings

Proposals Due: May 16, 2007, 2:00 P.M., local time

Fax Proposals to: 920-832-1534

E-mail Proposals to: purchamc@co.outagamie.wi.us

Delivery will be approximately ____________ days after receipt of purchase order.

Complete Attached Pricing Form

Firm Name: ______________________________
Authorized Signature: ______________________________
Print name: ______________________________
Title: ______________________________
Date: ______________________________
Address: ______________________________

City/State/Zip: ______________________________

Telephone: ______________________________
Fax: ______________________________

E-mail: ______________________________

Outagamie County East Landfill

HDPE Piping and Fittings Quantity Takeoff

Price Sheet

Due May 16, 2007

** All Items are SDR-7 HDPE unless noted otherwise.**

Item	Quantity	U/M	Unit Price	Extended Price
6-inch pipe	3000	ft		
4-inch pipe	2250	ft		
2-inch pipe	11900			
6 × 6 tee	12	ea		
4 × 4 tee	12	ea		
2 × 2 tee	48	ea		
6 × 4 Concentric reducers	12	ea		
4 × 2 Concentric reducers	20	ea		
2-inch 90 degree bends	60	ea		
2 × 1.5 Concentric Reducers	60	ea		
1.5 × 1 Concentric Reducers	60	ea		
1-inch HDPE × SS Transition Fittings (SDR-11)	60	ea		
1-inch Sch-40 PVC Temporary Caps for Transition Fittings	60	ea		
6 × 6 inch 90 degree bend	4	ea		
4 × 4 inch 90 degree bend	2	ea		

The Leon–Maxwell Performance Equation

In an attempt to simplify the product or service selection process, I have written an algorithm for performance value that can be assigned to a product or service—I call it the Leon–Maxwell Selection Standard. I named the equation after my father Leon and my son Maxwell.

This value, obtained from the Leon–Maxwell equation below (LM*n*), takes into account standards that best represent the performance of a product or service. The greater the number obtained, the better the product or service will perform and, in turn, save on downtime, labor costs, and replacement. This is especially helpful when taking into account many variables that influence overall performance.

The equation is as follows:

$$LMn = C \ln\left[\frac{\{A(Iv_1)^a \ldots Z(Iv_x)^z\}}{\{A(Dv_1)^a \ldots Z(Dv_x)^z\}}\right] \quad (5.1)$$

The spreadsheet version is now shown:

LM*n* = C *ln* {A(Iv1)^a)*...Z(Ivx)^z)} / {A(Dv1)^a)*...Z(Dvx)^z)}

The functional limitations of the standards or the physical limitations of the product or service determine the functional range for each value that has to be understood. The value of the constant C is used to provide a value base to 100 for a final value, which represents a product or service achieving the highest level of performance. Increasing performance values are found in the numerator I*v* and values that decrease as a function of increased performance are found in the denominator D*v*. The use of exponents a … z and multipliers A … Z normalize the results. The equation was derived in order to change the polynomial growth of the values into linear growth. Exponential growth curves are very difficult to distinguish. The use of the natural logarithm *ln* allows the relationship to become linear, which is easier to distinguish. This is a common technique used in the Richter and decibel scale measurements.

The derivation of the Leon–Maxwell equation consists of the following estimation of the original form of the Leon–Maxwell equation:

LM*n* = C *ln* {A(Iv1)^a)* … Z(Ivx)^z)}/{A(Dv1)^a)* … Z(Dvx)^z)} (5.2)

For simplicity, this can be rewritten as

$$\text{LMn} = \text{C}\ln\left\{\prod_{i=1}^{n} A_i Iv_i^{a_i}\right\} \quad (5.3)$$

For a term in the original denominator, the corresponding a_i will be negative. Taking the natural logarithm of the terms in the product results in

$$\begin{aligned} \text{LMn} &= \text{C}\sum_{i=1}^{n}\{\ln(A_i Iv_i^{a_i})\} \\ &= \text{C}\sum_{i=1}^{n}\{\ln(A_i) + a_i \ln(Iv_i)\} \\ &= \text{C}\sum_{i=1}^{n}\ln(A_i) + \text{C}\sum_{i=1}^{n} a_i \ln(Iv_i) \qquad (5.4) \\ LMn &= CA + \text{C}\sum_{i=1}^{n} a_i \ln(Iv_i) \\ \text{Where } A &= \sum_{i=1}^{n}\ln(A_i) \end{aligned}$$

That is, it is not the individual *A*s that are important, but the sum of their ln's. Equation (5.4) can be simplified further to

$$\begin{aligned} \text{LMn} &= \text{w}_0 + \sum_{i=1}^{n} w_i \ln(Iv_i) + \sum_{i=1}^{n} w_i \text{x}_i \\ \text{Where } \text{w}_0 &= \text{CA} = \text{C}\sum_{i=1}^{n}\ln(A_i) \qquad (5.5) \\ w_i &= Ca_i \\ x_i &= \ln(Iv_i) \end{aligned}$$

which is a linear equation in the weights w_i. If you have data relating LM*n* to the x_i's ($\ln Iv_i$'s), then the w_i's can be estimated using any linear regression package (such as Excel's Analysis Regression). The form of Equation (5.5) can be used to estimate other LM*n* values, or you can convert back to the C, A_i, and a_i constants. There is no unique conversion, as there are many extra degrees of freedom. For example, where C = 1 and $a_i = w_i$, then $A_i = \exp(w_0/n)$ is a valid conversion.

The selection standards are determined as they relate to the various performance standards. Many times, suppliers use ASTM (American Society for Testing and Materials) standards. ASTM has standards for just about any material or product and even some services. These are typically agreed upon industry standards. These standards provide the foundation for performance levels of many products.

In a functional example, a company wants to obtain the best steel for a fabrication project. They determined that the highest tensile strength (optimum is 10,000 psi) and the lowest elongation percentage (0.5) produces the best product. The equation developed would look something like this:

$$LMn = C\ Ln\ \{(\text{Tensile Strength in PSI})/(\text{Elongation }\%)\}$$

or

LM*n* =5*LN((Iv^2)/(Dv^2))

where 10,000 psi and 0.5 are the goals. The result would be an LM*n* of 100. The 100 term is the highest achievable value.

This is a simple example. The equation becomes more useful as the degree of complexity and amount of variables increase.

Application Examples

Under extreme conditions, ordinary conventional grease provides minimal protection. When examining the actual performance of various grease technologies and the relationship to certain ASTM values, the relationship becomes apparent. There are several indicators that point to grease performance in hostile conditions: bearing temperature, grease consumption, bearing failures, and energy requirements. The following are actual case studies of grease performance in various conditions.

Various standardized tests have been developed to provide for the determination and verification of performance characteristics. A short list of the ASTM tests for characterizing grease is presented in Table 5.7. When selecting grease for a specific application, the overall property of the grease should be considered. Table 5.7 describes the test method and how to interpret the test results. It is essential that the end user evaluate the data prior to using grease in a given application. The grease supplier should provide the customer with the desired test results (Table 5.3).

It has to be understood that these tests only provide a road map for final selection. To truly determine the best grease, a field study must be carried out. Choosing the best grease for a field trial can be difficult if not exhausting. When comparing the test data, it is helpful to develop a spreadsheet for easy comparisons. Table 5.7 is an example of such a comparison. There is a challenge in trying to understand which test or tests best mimic the actual environment that the grease will be experiencing. To make an accurate and informed decision on grease selection for any given application, a few tests stand out as good barometers of performance.

Table 5.3 Grease Test Methods and Result Values

Property	*Test Method*	*Description*	*Test Result Values*
Shear Stability	ASTM D 217	Multistroke penetration	The lower the percent change in the number, the more mechanically stable the grease.
	ASTM D 1831	Roll stability	The lower the percent change in the number, the more mechanically stable the grease.
	ASTM D 1263	Wheel bearing leakage	Measures percent loss in a wheel-bearing application. The lower the number the better, above 5% will cause brake problems.
Oxidation Resistance	ASTM D 942	Bomb oxidation	Measures the oxidative life of the grease. The lower the percentage, the better the oxidation resistance.
	ASTM D 3527	Wheel bearing life	The higher the hours, the longer the grease will last in high-temperature applications.
	ASTM D 3336	High-temperature performance	The higher the temperature, the better the grease will perform at high temperatures.
Water Resistance	ASTM D 1264	Water washout	The lower the percent, the less likely it will wash out.
	ASTM D 4049	Water spray-off	The lower the percent, the less likely it will wash out.
Bleed Resistance	FTM 321.3	Oil separation (static)	Measures the percent oil that may separate during storage and idle time.
	ASTM D 1742	Pressure oil separation	Measures percent oil that will separate when grease is under load.

Extreme-Pressure/ Antiwear	ASTM D 2596	Four-ball	Point contact, similar to ball bearings; the higher the number, the greater load carrying.
	ASTM D 2509	Timken method	Line contact, similar to roller bearings; the higher the number, the greater load carrying.
	ASTM D 2266	Four-ball (wear scar)	The lower the number, the more protection.
Corrosion	ASTM D 1743	Rust test	Determines how well the grease keeps water and corrosives away from the metal surface, static test.
	—	Emcor	Determines how well the grease keeps water and corrosives away from the metal surface, dynamic test.
	ASTM D 130	Copper corrosion	1A is the best rating, most are 1B, measures ability to protect yellow metals.
Pumpability	ASTM D 4693	Low-temperature torque	Measures the effort required to move the grease in a bearing at low temperatures. The lower the number, the better.
	US Steel LT37	Mobility	Measures the grease flow at a given temperature at 150 psig. The higher the number, the better; critical is 2 grams per minute.
Identification and Quality Control	ASTM D 2265	Dropping point	Measure the temperature the soap melts, used to help determine the upper usable temperature range.

For the lubricating grease example, the Leon–Maxwell equation would be

$$LMn = (4.9615)\ln\left[\frac{\{(D_W)^2(P_D)(L_S)\}(0.001)}{(R_W)(W_W^{1.5})\{(Ox+2)^2\}(P_W+1)}\right] \tag{5.7}$$

where

D_w = ASTM D-2596 four-ball weld value in kg (functional range from 200 to 1000).

P_D = ASTM D-2265 dropping point in °F (functional range from 250 to 600, nonmelt grease assumes a value of 600).

L_S = percent lubricating solids, for example, molybdenum disulfide, graphite, Teflon, calcium sulfonate, percent expressed in whole numbers (functional range from 1 to 100).

R_w = ASTM D-2596 four-ball wear scar in mm (functional range from 0.01 to 1).

W_w = ASTM D-1264 water washout percent expressed in whole numbers (functional range from 0.1 to 100).

Ox = ASTM D 942 bomb oxidation pressure loss expressed in whole numbers (functional range from 0 to 100).

P_W = ASTM D 217 penetration, worked 60 to 100,000, percent change expressed in whole numbers (functional range from 0 to 100).

The following is an example of the Leon–Maxwell equation and in spreadsheet form in Table 5.4 using Equation (5.2):

$$\text{LM}n = 4.9615\text{*LN}(((D_W \text{ ^}2\text{*}(P_D)\text{* } L_S)\text{*}(0.001))/(R_W \text{ *}(W_W \text{ ^}1.5)\text{*} (O_X +2)\text{^}2\text{*}(P_W +1)))$$

Excessive Load Example

An iron forging plant utilizes a 50,000+ pound flywheel mounted on two 500-mm Torrington roller bearings. The flywheel rotates at 300 rpm. The operating temperatures are held in strict control not to exceed 218°F. Thermisters were retrofitted into the bearing housing to closely monitor the operating temperatures. These thermisters were routinely calibrated and tested for accuracy. It was calculated that if the bearing race was to exceed 218°F, the rollers would swell and shrink the dimensional tolerance of the bearing. Once the tolerance was compromised, the grease that was being used would then be forced out and metal-to-metal contact would ensue. Once this began, the operational temperature of the bearing would rise dramatically, and possible seizure could take place. Replacing the bearing would cost in excess of $750,000 worth of parts and labor. Because of the location, if the flywheel was to seize, a high-capacity boom crane would have to be brought

Table 5.4 Leon–Maxwell Numbers Calculated Using a Spreadsheet

Four-Ball Weld	*Dropping Point*	*Percent Solids*	*Four-Ball Wear*	*Water Washout*	*Bomb Oxidation*	*Worked Penetration*	*Leon–Maxwell Number*
D_W	P_D	L_S	R_W	W_W	O_X	P_W	LM*n*
1000	**600**	**40**	**0.341**	**0.1**	**0**	**0**	**100**
800	500	20	0.32	0.1	1	0	**90**
650	550	25	0.4	0.5	1	2	**71**
500	500	20	0.35	0.75	1	5	**61**
300	450	10	0.5	2	0	0	**55**
300	600	20	0.8	0.5	1	5	**55**
450	400	10	0.5	3	0	1	**52**
200	350	10	0.38	0.5	1	2	**52**
500	350	3	0.5	1	3	2	**44**

in and the roof removed. The flywheel was typically in operation for only 2 hours per shift before the temperature would rise to unacceptable levels. The grease used in the bearings was a nonmelt bentonite with a synthetic base oil. The decision was made to switch to a higher load-carrying grease in order to achieve longer run times. It was theorized that the load-carrying ability of the grease was not high enough and that the mechanical stability of the grease was being compromised. A high solids calcium sulfonate grease was selected according to the four-ball weld and wear values and worked penetration stability. After the changeover, the flywheel runtime was extended to 8 hours and grease consumption was reduced by one-fourth (Graphs 5.1 and 5.2). The four-ball weld and wear values, lubricating solids percentage and mechanical stability, and Leon–Maxwell numbers (LM*n*) are compared in Table 5.5. The higher LM*n* performs the best.

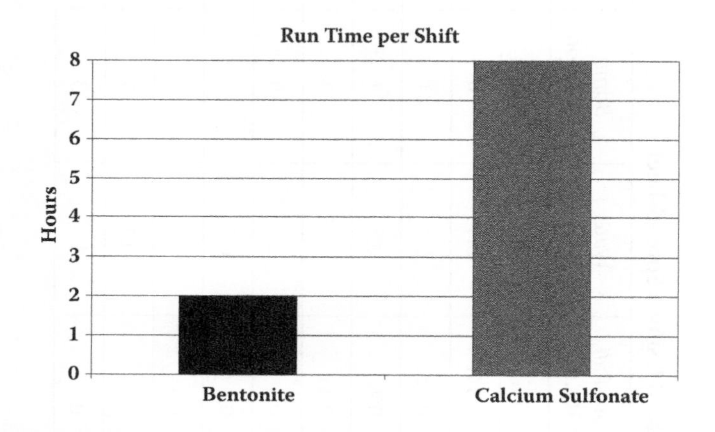

Graph 5.1 Run-time comparison.

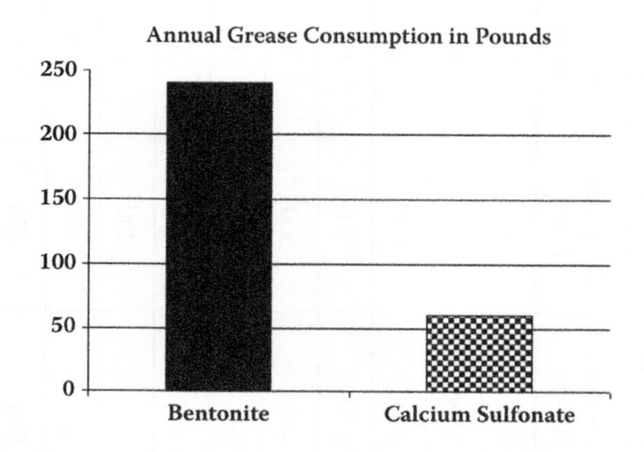

Graph 5.2 Grease consumption.

Table 5.5 Calcium Sulfonate Compared to Synthetic Bentonite for Load

Failure Mode	*Selected ASTM Standard*	*Synthetic Bentonite*	*Calcium Sulfonate*
Lack of extreme load handling of grease	Four-Ball Weld and Wear	450 0.75	1000 0.38
Lack of mechanical stability of grease	Worked Penetration Percent Change	8	<1
	Leon–Maxwell Numbers (LM*n*)	68	95+

Excessive or Localized Heat

A brick manufacturing plant was experiencing lubrication failures in the wheel bearings in their kiln cars. Kiln cars operate in extremely heavy load, high heat, and heavily contaminated conditions. Each car has eight wheel bearings to support the load of the tens of thousands of pounds placed on it. When a wheel bearing fails, the wheel locks up, and the kiln car is pushed through the remainder of the process. In addition to requiring more energy to push the car, the wheel, the shaft, and the bearing all have to be rebuilt or replaced. It is not uncommon for the wheel bearings to experience continuous temperatures of 800°F to 1200°F for up to 6 days. The grease being used was a nonmelt bentonite thickened grease with a synthetic base oil. Upon inspection of the failed bearings, it was apparent that the grease being used was forming abrasive solids at high temperatures and that the grease was showing signs of severe oxidation. The decision was made to switch four cars over to a higher load-carrying grease in order to achieve reduced bearing failures. The cars were fitted with existing bearings packed with the new grease. It was theorized that the load carrying ability of the grease was not high enough and that the oxidation stability of the grease was being compromised. High solids calcium sulfonate grease was selected according to four-ball weld and wear values and oxidation stability. After the changeover, the cars were closely monitored and compared to existing cars. Initial results suggested that the new grease was performing much better than the previous grease. Complete change-over was made, and grease consumption and parts replacement were closely monitored. The maintenance staff continued to lubricate according to their established schedule. It was noticed that the amount of new grease added to the bearings was much less than previously used. The consumption was reduced 54%. Parts replacement was reduced dramatically. The new grease was simply staying in the bearings longer and providing more protection. Downtime and grease consumption are compared in Graphs 5.3 and 5.4. The four-ball weld and wear values, lubricating

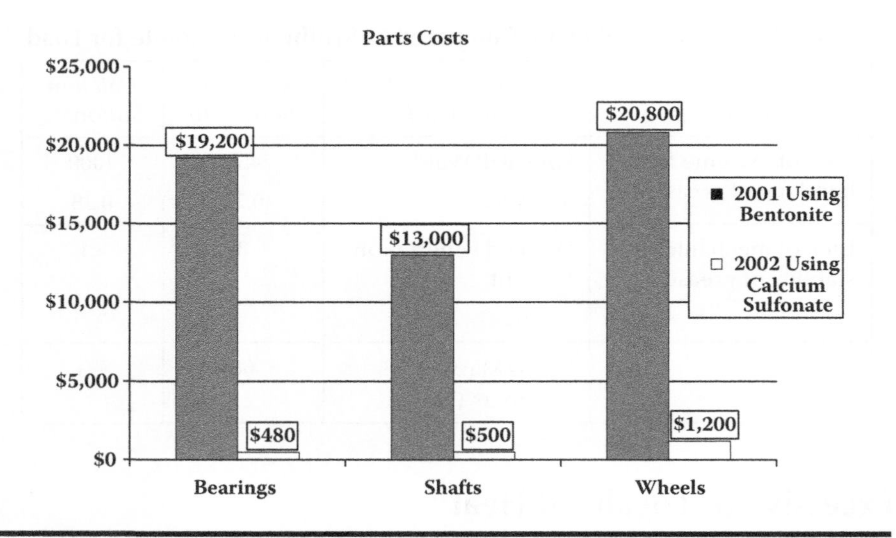

Graph 5.3 Parts costs.

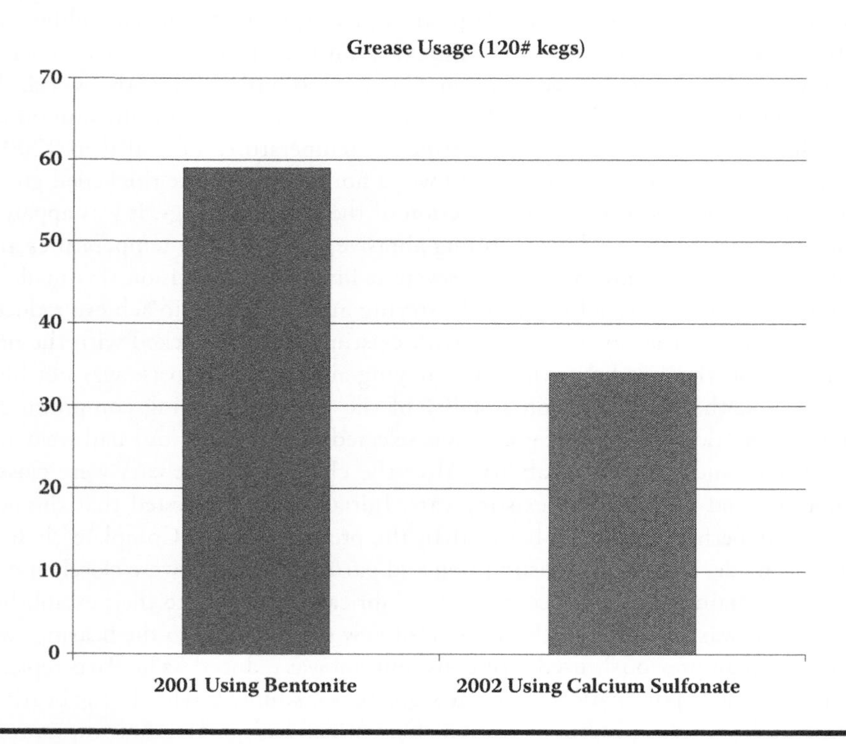

Graph 5.4 Grease usage.

Table 5.6 Calcium Sulfonate Compared to Bentonite for Oxidation

Failure Mode	*Selected ASTM Standard*	*Synthetic Bentonite*	*Calcium Sulfonate*
Lack of extreme load handling of grease	Four-Ball Weld	380	1000
Lack of oxidation resistance of the grease	Bomb Oxidation	12	<1
	Leon–Maxwell Number (LM*n*)	61	95+

solids percentage and mechanical stability values, and Leon–Maxwell numbers are compared in Table 5.6. The higher LM*n* performs the best. Using grease with a high LM*n* saved over $60,000.

Contamination

A protein rendering plant, which uses large cookers, operates at high temperatures. The cookers have 420-mm bearings and have experienced hardened grease resulting in frequent bearing failures. The process produces contaminants—solid, steam, and water—that enter bearings and also contribute to bearing failures. The cookers operate continuously, and the energy consumption is closely monitored. The operating temperature of the unit is 320°F. The bearings are radiant heated and stay at elevated temperatures continuously due to the nature of the process. A lithium complex thickened grease was being used with minimal success. Upon inspecting the failed bearings, it was noticed that many of the bearings appeared dry or, in some cases, the grease appeared to have hardened. The decision was made to switch over to higher oxidation resistant and mechanically stable grease in order to achieve reduced bearing failures. A high solids calcium sulfonate grease was selected. After the introduction of the grease into the bearings, the energy consumption dropped 30% and the grease consumption was reduced 75% (Graphs 5.5 and 5.6). To date, bearing failures related to grease have been eliminated. The four-ball weld and wear values, oxidation resistance, mechanical stability values, and Leon–Maxwell numbers are compared in Table 5.7. The higher LM*n* performs the best. Using grease with a high LM*n* saved over $120,000.

The Leon–Maxwell equation provides an excellent indicator for selecting products or services. The Leon–Maxwell equation is a concise equation that takes into consideration several test results or standards to establish a single numeric value for selection and consolidation purposes on a scale to 100. By using a product or

Table 5.7 Calcium Sulfonate Compared to Lithium Complex for Load

Failure Mode	*Selected ASTM Standard*	*Lithium Complex*	*Calcium Sulfonate*
Lack of extreme load handling of grease	Four-Ball Weld and Wear	400 0.63	1000 0.38
Lack of oxidation resistance of the grease	Bomb Oxidation	5	<1
Lack of mechanical stability of grease	Worked Penetration Percent Change	7	<1
	Leon–Maxwell Number (LM*n*)	55	95+

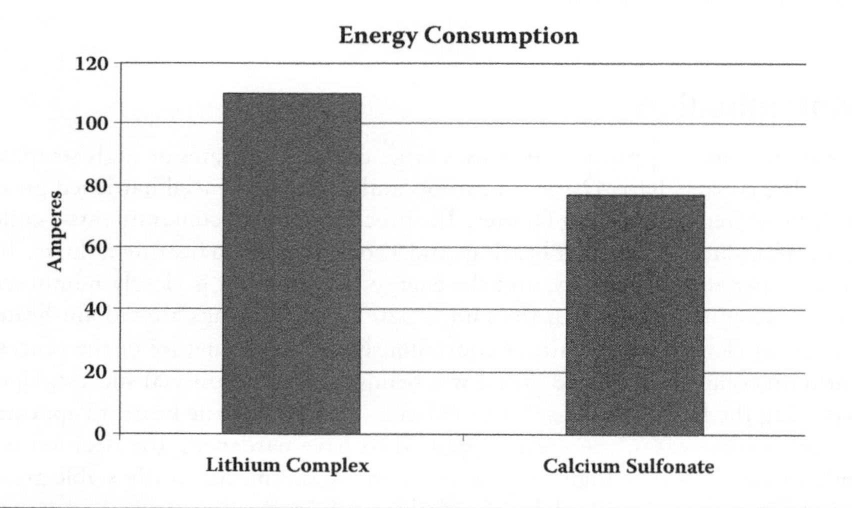

Graph 5.5 Energy consumption.

service with the highest Leon–Maxwell number, increased reliability and reduced replacement costs can be realized.

The evaluation process is deciding on what you are trying to achieve and why. These tools are also helpful in determining the best product. There are several tools that have proven to be very helpful in the evaluation process. Each has their strengths and pitfalls. Your comfort level with diagramming, spreadsheets, or interviewing will dictate which method will be most useful. What you are essentially trying to accomplish is to be able to determine what the best product to purchase is that will have an effect on the overall operating costs.

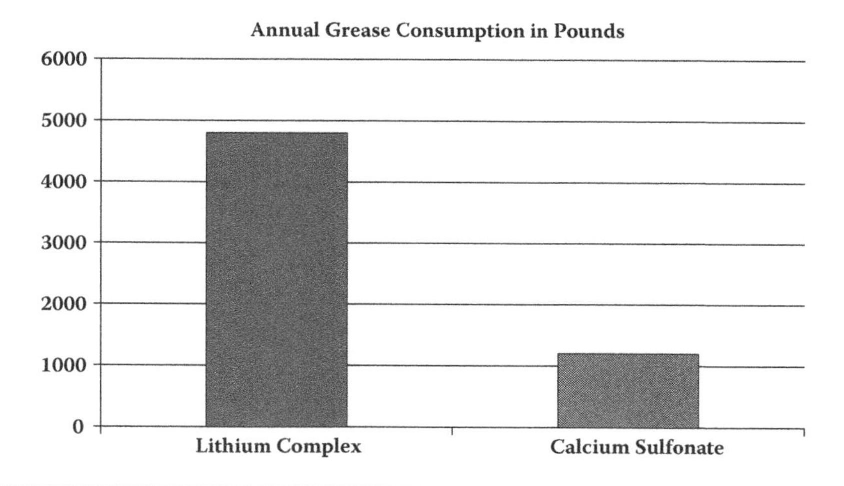

Graph 5.6 Annual grease consumption.

The Leon–Maxwell equation can be a powerful tool to help determine which response is best suited for your particular need. The following process can aid in the property and standard determination:

1. Rank what is required.
2. Translate the action—from "what" to "how." An example would be adhesion of a glue to the peel strength of the glue.
3. Determine relationships and correlations—values must be assigned to the relationships as well as the direction of importance.

In Table 5.8, values will be assigned according to the correlation of the requirements and the properties.

The process is as follows:

1. List the requirements.
2. Rank each requirement according to importance.
3. List the properties.
4. Correlate the properties to the requirements using a three-point scale with 3 = strong correlation, 2 = medium correlation, and 1 = weak correlation.
5. Cross-multiply the correlation to the rank and prioritize through a summation.
6. Establish the risk priority number (RPN) for the product. It is the Severity (S), Occurrence (O), and Detection (D) ranking:

$$\text{RPN} = (S) \times (O) \times (D)$$

Table 5.8 Performance Ranking

		Characteristics or Physical Performance/ Property		
Requirements	*Ranking of Importance*	*Property 1*	*Property 2*	*Property 3*
Requirement 1				
Requirement 2				
Requirement 3				
Requirement 4				
Requirement 5				
		Standard 1	Standard 2	Standard 3
		How Characteristics or Physical Performance/ Property Are Measured (Standards)		

where severity (S) is an assessment of the seriousness of the effect of the potential failure mode. Severity applies to the effect only. A reduction in severity ranking index can be effected only through a design change. Severity is estimated on a 1 to 10 scale. An example of severity ranking can be found in Table 5.9.

Occurrence (O) is the likelihood or probability that a specific cause will occur. Occurrence is estimated on a "1" to "10" scale (Table 5.10).

Detection (D) is an assessment of the ability of proposed current design controls (where the cause/mechanism or failure mode/effect from occurring, or reduce their rate of occurrence) to detect a potential cause/mechanism (design weakness), or the ability of the proposed current design controls to detect the subsequent failure mode, before the component, subsystem, or system is released for production. To achieve a lower ranking, generally the planned design control (e.g., preventative, validation, and/or verification activities) has to be improved. Detection (D) is estimated on a "1" to "10" scale (see Table 5.11).

The RPN is a measure of *design risk*. The RPN is also used to rank order the *concerns in processes* (e.g., in Pareto fashion). The RPN will be between 1 and 1000. For higher RPNs an effort must be made to reduce this calculated risk through corrective action(s). When the failure modes have been rank ordered by RPN, corrective action should be first directed at the highest ranked concerns and critical items. The intent of any recommended action is to reduce any one or all of the occurrence, severity, and detection rankings.

After the corrective action(s) have been identified and the specification in place, you should then estimate and record the resulting severity, occurrence, and detection rankings. Calculate and record the resulting RPN. If no action is

Table 5.9 Severity Ranking

Effect	*Criteria: Severity of Effect*	*Ranking*
Hazardous—Without Warning	Very high severity ranking when a potential failure mode affects safe operation or involves noncompliance with regulations without warning	10
Hazardous—With Warning	Very high severity ranking when a potential failure mode affects safe operation or involves noncompliance with regulations with warning	9
Very High	Product/item inoperable, with loss of primary function	8
High	Product/item operable, but at reduced level of performance, customer dissatisfied	7
Moderate	Product/item operable, but may cause rework/repair or damage to equipment	6
Low	Product/item operable, but may cause slight inconvenience to related operations	5
Very Low	Product/item operable, but possesses some defects (aesthetic and otherwise) noticeable to most customers	4
Minor	Product/item operable, but may possess some defects noticeable by discriminating customers	3
Very Minor	Product/item operable, but is in noncompliance with company policy	2
None	No effect	1

taken, leave the "Resulting RPN" and related ranking columns blank. All resulting RPNs should be reviewed, and if further action is considered necessary, then repeat your steps.

Green Procurement

One area that has taken the country by storm has been the Green Movement. Elizabeth Rogers and Thomas Kostigen wrote a book entitled *The Green Book*. The message it sends contends with the simple fact that if we choose to change several habits around the house, we can have a dramatic difference in the environment. From refusing ATM receipts, to turning off the water while we brush our teeth, and even using dishwasher-safe flatware as opposed to fine china that requires

Table 5.10 Occurrence Ranking

Probability of Failure	*Possible Failure Rates*	*Ranking*
Very High: Failure is almost inevitable	≤1 in 2	10
	1 in 3	9
High: Repeated failures	1 in 8	8
	1 in 20	7
Moderate: Occasional failures	1 in 80	6
	1 in 400	5
	1 in 2000	4
Low: Relatively few failures	1 in 15,000	3
	1 in 150,000	2
Remote: Failure is unlikely	≥1 in 1,500,000	1

hand washing—we could have an impact on our ecology and our environment. Although some of their assertions require number crunching, the overall message is sound—we must become more responsible and consider the total ramifications of our habits, including purchasing habits.

A serious consideration of many companies and agencies is "green buying." This refers to the procurement of materials and items that help reduce pollution and energy demands. Green buying is buying environmentally responsible or "green" products and services that minimize environmental impacts. This means looking at the environmental consequences of a product at all the various stages of its life cycle—from processing raw materials, manufacturing, transporting, storing, handling, using, to disposing of the product.

Green products are produced using fewer natural resources or using them with an emphasis on sustainability. They also may involve less energy in production, consume less energy when being used, and contain fewer hazardous or toxic materials. A product can be green if it helps reduce the amount of waste created. Green procurement can also offer cost savings through buying products that are easily recycled, last longer, or produce less waste. Green is more than the color of a tree being saved; it is the color of money as well. Money is saved on waste disposal and as a result, requires fewer resources to manufacture and operate, so savings can be made on energy, water, fuel, and other natural resources. Green products have fewer toxic or hazardous materials. This reduces expenses such as permit fees, toxic materials handling charges, and staff training.

Table 5.11 Detection Ranking

Detection	*Criteria: Likelihood of Detection by Design Control*	*Ranking*
Absolute Uncertainty	Design control will not or cannot detect a potential cause/mechanism and subsequent failure mode; or there is no design control	10
Very Remote	Very remote chance the design control will detect a potential cause/ mechanism and subsequent failure mode	9
Remote	Remote chance the design control will detect a potential cause/mechanism and subsequent failure mode	8
Very Low	Very low chance the design control will detect a potential cause/ mechanism and subsequent failure mode	7
Low	Low chance the design control will detect a potential cause/mechanism and subsequent failure mode	6
Moderate	Moderate chance the design control will detect a potential cause/ mechanism and subsequent failure mode	5
Moderately High	Moderately high chance the design control will detect a potential cause/ mechanism and subsequent failure mode	4
High	High chance the design control will detect a potential cause/mechanism and subsequent failure mode	3
Very High	Very high chance the design control will detect a potential cause/ mechanism and subsequent failure mode	2
Almost Certain	Design control will almost certainly detect a potential cause/mechanism and subsequent failure mode	1

Many companies will require a green procurement program as part of an ISO 14001 policy. The definition is as follows.

The ISO 14001 Standard

> It is a fundamental principle of ISO 14001, which governs environmental management systems that organizations set their own goals, based on whatever considerations they wish to include, such as the demands of customers, regulators, communities, lenders or environmental groups. The ISO 14001 standard provides a framework within which to develop plans to meet those targets, and to produce information about whether or not the targets are met. By the end of 2001, nearly 32,000 organizations worldwide had received ISO 14001 accreditation. An important benefit of adopting ISO 14001 is to give stakeholders the reassurance they need that the organization's environmental claims are valid.
>
> The ISO 14001 standards are intended to be flexible, and to be of value in a wide variety of situations. However, it is applicable most readily to large companies that already have a formal management system in place, and which have the expertise and resources to incorporate environmental issues into that system. However, the principles have been designed to apply also to smaller businesses, and to non-business organizations.

In general, conformance with one of the international standards can lead to the implementation of other standards. For example, once an organization has implemented ISO 14001, it is easier for it to satisfy the requirements of EMAS (the European Eco-Management and Audit Scheme).

Companies and agencies that engage in green procurement are considered to be socially responsible. This is important for an organization wanting to have "Corporate Social Responsibility," which is also called CSR, corporate responsibility, corporate citizenship, responsible business, and corporate social opportunity. This approach is a concept for which organizations consider the interests of society by taking responsibility for the impact of their activities on customers, suppliers, employees, shareholders, communities, and the environment. It is a huge selling and marketing feature as well as the next buzz.

The pitfalls of green procurement include:

Price: Sometimes a nontoxic formula will be more expensive due to lower market demands and less volume.

Lack of Commitment: All levels of the organization must be involved. This can be very difficult to police if there is no policy in place.

Lack of Knowledge: The concept is relatively new and many do not know various terms, concepts, or considerations.

No Acceptable Alternatives: Some "nongreen" materials or items perform better than the greener versions, and in some instances, there is not a green version available.
Lack of Specifications: Many specifications do not include green requirements such as toxicity or waste considerations.

The following are steps involved in implementing a green procurement program:

1. *Organizational Support*: Establish policies and procedures. It is essential that management support the initiative fully. In addition, those charged with making purchasing decisions must be involved in the implementation process.
2. *Self-evaluation*: Audit purchasing practices. This process will help to clarify what is purchased, in what quantities, from where, and at what price, and compare to the goals set forth in step 1.
3. *Establish a Policy*: Should include specific priorities and targets.
4. *Develop Short- and Long-Term Strategy*: Identify suitable products and services, and evaluate the environmental performance of suppliers.
5. *Run a Case Study*: Apply green procurement principles to a specific product or service. A case study can be used to generate more detailed guidance on purchasing practices while understanding and documenting the overall cost savings and effect.
6. *Implementation*: By incorporating departmental requirements as well as safety, environmental, and management goals.
7. *Sustain*: A systematic review of the process has to be carried out on a periodic basis in order to establish whether the scheme is meeting its goals and objectives. The review should take into account changing environmental goals.

The Department of Defense (DoD) has set up the following Green Procurement Program (GPP) objectives:

- Educate all appropriate DoD employees on the requirements of federal "green" procurement preference programs, their roles and responsibilities relevant to these programs, and the DoD GPP, and the opportunities to purchase green products and services.
- Increase purchases of green products and services consistent with the demands of mission, efficiency, and cost-effectiveness, with continual improvement toward federally established procurement goals.
- Reduce the amount of solid waste generated.
- Reduce consumption of energy and natural resources.
- Expand markets for green products and services. The DoD is committed to becoming a leader in green procurement. As the single largest buyer of supplies and services throughout the government, DoD must strive to ensure

that every procurement meets the requirements of all applicable federal green procurement preference programs.

Further information can be found at the following:

Buying Green: "A Multi Functional Approach to Pollution Prevention." To order the manual or obtain more information on training offered by the DLA Training Center, call (614) 692-5969, 1-800-458-7903, (269) 961-7046, or fax (269) 961-7055.

Green Procurement Seminar: The U.S. Army Center for Health Promotion and Preventive Medicine (USACHPPM) offers onsite Green Procurement training to credit card buyers, procurement request originators, and contracting personnel (http://chppm-www.apgea.army.mil/gwswp/SolidWaste/training.htm).

Online Green Purchasing Training: "What is Green Purchasing, Anyway?" is an online green purchasing training course for contracting personnel, purchase card holders, facility personnel, and product specifiers. Designed in modules, it can be used both for introductory training and for refresher training. It is available on the Office of Personnel Management's GoLearn e-learning center (www.golearn.gov). Affirmative Procurement: DAU Contracting Courses (http://www.dau.mil/).

The following is a list of tools and guidance documents available online:

- For a quick reference guide to various programs and products involved in DoD's GPP, see the AFCEE GPP Fact Sheet at: http://www.afcee.brooks.af.mil/eq/ap/GPP_Factsheet.doc.
- For product and specification information, visit the U.S. Environmental Protection Agency (EPA) database at: http://www.epa.gov; and follow the links to see detailed explanations of products, guidance, and supply sources.
- The "Guide to Green Purchasing," which was updated in January 2005 to reflect the full scope of the GPP, can be downloaded by accessing Guide to Green Purchasing.
- For information regarding GPP-Compliant Recycled Content Products, visit the EPA Web site for updated CPG product lists at: http://www.epa.gov/cpg/products.htm.
- For lists of products and manufacturers that have earned the Energy Star® seal, visit: http://www.energystar.gov/products.
- For Low Standby Power devices and manufacturer lists, visit: http://oahu.lbl.gov.
- For alternatives to ozone-depleting substances, visit: http://www.epa.gov/ozone/snap/lists/index.html.

- For environmentally preferable purchasing, visit: http://www.epa.gov/epp.
- For energy-efficient products, visit: http://www.eere.energy.gov/femp/procurement/.
- For biobased products, visit: www.biobased.oce.usda.gov and http://www.ofee.gov.
- For alternative fuels and fuel efficiency, visit: http://www.eere.energy.gov/vehiclesandfuels/.
- For nonozone-depleting substances, visit: http://www.ofee.gov.
- For a list of green procurement definitions, contact PROACT at DSN 240-4240 or e-mail pro-act@brooks.af.mil.

We often see various logos and seals denoting the green effort. The following is a primer on the various entities:

Green Seal: The top 15% to 20% of products in a category can earn a *Green Seal* label. Green Seal considers several factors: level of volatile organic compounds (VOCs) (fewer than 150 grams of VOCs per liter) of product, minus water. No hazardous chemical compounds were used in manufacturing (benzene, naphthalene, and mercury, etc.)—none can be present. The packaging must discourage consumers from improper disposal. The performance level is not diminished in comparison.

Greenguard: Focuses on indoor air quality, off-gassing products, and products that have an impact on indoor air and the building's occupants.

Design for the Environment (DfE): A voluntary partnership program sponsored by the EPA, which screens each ingredient for potential human health and environmental effects and ensures that only those ingredients that pose the least concern among chemicals in their class are present.

EcoLogo: Canada's multiattribute green certification program with over 120 product categories, developed on the life cycle of a product. The criteria are similar to Green Seal.

LEED® Program (Leadership in Energy and Environmental Design): Developed by the U.S. Green Building Council, it is the most recognized and adopted measurement tool for green buildings. The program is a voluntary, consensus-based rating system for developing high-performance, sustainable buildings. Ratings systems are available for new construction or existing buildings.

Conclusion and Summary

After you have received all responses, it is now time to rank them. This is a very easy exercise. Basically, by listing out the total LM*n* for each, it becomes obvious which suppliers to go with. It is highly suggested that the top two or three are chosen.

When done so, select areas to carry out a case study, as described earlier. Once success is understood, then sweeping changes can take place.

At this point, all the stages of developing a performance-based purchasing specification have been explained. The degree of complexity to which you choose to establish a specification is entirely up to your energy level. Keep in mind, it would serve your best interest to keep focused on the products, properties, and standards that best suit the given situation. If the specification is complicated, it may become a burden for legitimate suppliers to respond. The following is a checklist for the various tasks that should be carried out. This is only a suggestion of the process. You may have to adjust, add, or delete various stages according to your departmental goals and your company vision. It must always be understood that the purchasing specification is put in place in order to help drive down operational costs. It is must be considered a living document and be changed according to the changing conditions of production.

Specification Development Checklist

	Checklist Item	*Y*	*N*	*N/A*	*Comments*
1	Adequate and appropriate records are maintained throughout the procurement process and provide sufficient information to enable an audit or independent review.	☐	☐	☐	

Specification Planning

	Checklist Item	*Y*	*N*	*N/A*	*Comments*
2	Appropriate approval to undertake purchase has been obtained in accordance with company requirements, mandates, and vision.	☐	☐	☐	
3	Appropriate procedures are in place to ensure that persons submitting responses to the specification requirements are dealt with fairly and equitably during the tender process.	☐	☐	☐	
4	Product—Tool/component/material is identified through Collection Management Mission Applications (CMMA), interviews.	☐	☐	☐	

Specification Planning (Continued)

	Checklist Item	*Y*	*N*	*N/A*	*Comments*
5	Product—Tool/component/material cost of failure in terms of downtime, labor burden, parts replacement has been established through records, CMMA, interviews.	☐	☐	☐	
6	Product—Tool/component/material performance criteria have been established.	☐	☐	☐	
7	Product—Tool/component/material performance criteria physical tests and standards have been identified that reflect required performance and quality levels.	☐	☐	☐	

Preparing the Specification

	Checklist Item	*Y*	*N*	*N/A*	*Comments*
8	The purchasing specification provides all the information necessary to enable potential suppliers to prepare appropriate submissions.	☐	☐	☐	
9	(a) A clear description of the goods or services to be procured.	☐	☐	☐	
10	(b) All conditions for participation.	☐	☐	☐	
11	(c) Details of the evaluation criteria to be used in the assessment of tenders, the evaluation methodology, and any weightings to be used in the assessment.	☐	☐	☐	
12	(d) Details of the information that should be provided by suppliers.	☐	☐	☐	
13	(e) All other relevant terms and conditions of the tender.	☐	☐	☐	
14	(f) Details of any applicable policies and principles.	☐	☐	☐	
15	(g) Details of the contact information.	☐	☐	☐	

(*Continued*)

Preparing the Specification (Continued)

	Checklist Item	*Y*	*N*	*N/A*	*Comments*
16	(h) Details of the closing date.	☐	☐	☐	
17	(i) Advice on the treatment of late responses.	☐	☐	☐	
18	(j) Pricing requirements (e.g., price to be exclusive of goods or service tax).	☐	☐	☐	
19	(k) Indication as to whether alternative responses will be considered.	☐	☐	☐	

Receiving Responses

	Checklist Item	*Y*	*N*	*N/A*	*Comments*
20	Fair and impartial procedures were in place in relation to the opening of the responses.	☐	☐	☐	
21	Potential suppliers have been provided with an opportunity to correct unintentional errors of form between the opening of submissions and any decision. The same opportunity was provided to all participating potential suppliers.	☐	☐	☐	
22	Respondents were advised that their submissions were received.	☐	☐	☐	
23	Information provided by persons submitting replies is treated as confidential.	☐	☐	☐	
24	Documents have been secured.	☐	☐	☐	
25	The specification has been reviewed by the legal department to ensure that it does not break any trade or purchasing laws.	☐	☐	☐	
26	The specification has been reviewed by the production department to ensure that it is in compliance with operational procedures.	☐	☐	☐	

Receiving Responses (Continued)

	Checklist Item	*Y*	*N*	*N/A*	*Comments*
27	The specification has been reviewed by the maintenance department to ensure that it meets performance and quality objectives for increasing equipment reliability and reducing downtime, parts replacement, and labor burden.	☐	☐	☐	
28	A specification evaluation plan has been developed.	☐	☐	☐	
29	An evaluation committee has been established, and members are familiar with procurement processes.	☐	☐	☐	
30	Evaluation criteria, weighting, and an evaluation methodology have been defined.	☐	☐	☐	
31	The evaluation criteria have been based on the specification values and weighted accordingly.	☐	☐	☐	

Evaluation

	Checklist Item	*Y*	*N*	*N/A*	*Comments*
32	Responses are fairly and equitably evaluated in a manner that is consistent with the Fair Trade Act as well as corporate guidelines.	☐	☐	☐	
33	The evaluation criteria, weightings, and methodology as set out in the specification have been used to evaluate the respondents.	☐	☐	☐	
34	The reasons for not accepting any response have been documented on file and are clear and justifiable (e.g., substantially nonconforming; specified quality assurance requirements not met; insufficient expertise or inadequate filing procedures).	☐	☐	☐	

(*Continued*)

Evaluation (Continued)

	Checklist Item	*Y*	*N*	*N/A*	*Comments*
35	The sale (and potential contract) is being awarded to the supplier who:				
36	(a) Satisfies the conditions for participation; and	☐	☐	☐	
37	(b) Is fully capable of undertaking the task and burdens; and	☐	☐	☐	
38	(c) Whose submission is determined to be the best value, or the most advantageous in accordance with the essential requirements and evaluation criteria specified in the documentation.	☐	☐	☐	
39	A specification respondent evaluation report has been completed and signed by all members of the evaluation committee.	☐	☐	☐	
40	Confirmation has been sought regarding the availability of funds for the actual cost of the goods/services.	☐	☐	☐	

Notifications

	Checklist Item	*Y*	*N*	*N/A*	*Comments*
41	A submission was made to the review committee, using the appropriate forms, seeking endorsement of the procurement process.	☐	☐	☐	
42	The review committee endorsed the process used in the procurement.	☐	☐	☐	
43	The recommendation of the evaluation committee has been approved by the appropriate delegated authority.	☐	☐	☐	
44	The successful and unsuccessful suppliers have been advised of the outcome of their responses.	☐	☐	☐	

Appendix

The Cover Letter

The *request for proposal* (RFP) *cover letter* should accompany the RFP questionnaire. Why write a cover letter? Beyond being polite and presenting your project, the RFP cover letter gives you a unique opportunity to emphasize the timeline of your state-of-the-art RFP-based selection process, particularly the dates on which different documents are due.

It is highly recommended that you read the suggestions below in order to properly and successfully use the RFP cover letter:

1. Use a formal letterhead, and do not handwrite the RFP cover letter.
2. First, *invite the provider to submit a proposal* based on the requirements defined in the RFP document you will attach to the cover letter.
3. Next, *present your company* with a brief description of your principal business, objectives, and location, and a link to your Web site.
4. Then, *describe your existing systems*, your project (size, budget, scope, etc.), and how you want the project to be aligned with your business objectives.
5. *Present your project timeline.* The success of your process may depend on properly communicating both the critical steps of your process and the deadlines the providers should meet in order to have their proposal considered. Missed due dates are the main cause of delays. Therefore, the more emphasis you put on deadlines, the shorter your process will be.

Highlight the major events and the due dates for submitting information:

- The *preproposal meeting*. Decide whether or not to make it mandatory.
- *Reservation* of spots for the preproposal meeting. Decide whether or not vendors must reserve spots for the preproposal meeting.

- *Questions* for the preproposal meeting. For the sake of convenience, you may decide to publish questions and answers on your Web site. Providers may also be able to submit their questions online.
- The *letter of intent*. A model of a letter of intent should be provided in the RFP. Decide if you want the letter of intent to be mandatory or to merely help you manage the process.
- The *proposal*. Specify your requirements (number of copies, structure, formats, electronic versions, etc.).
- The *contract award* of the project to the most satisfactory proposal. It is not necessary to commit to a firm date, but an estimated time frame could be provided.

Regardless of the types of responses you receive, providing a clear, precise timeline will ensure that your selection process is on time and within budget. In fact, by making certain you get the right documents in a timely manner, you will avoid undue delays, and thus save money.

When giving a deadline, be aware of the noninterchangeability of the words "submit" and "receive." More often than not, RFP authors write "*submit* the document no later than," but expect to *receive* the document no later than that date, thus generating proposal disqualifications, protests, and significant delays.

Again—it cannot be said enough—make sure prospective providers understand which documents to submit, receive detailed and precise instructions regarding the format of these documents, and are aware of your deadlines for receiving them. A consistent format for their proposal will be very helpful in evaluating all responses in a faster and more precise manner.

1. Thank the person for his or her time, effort, and interest in the project related to the issued RFP.
2. Finally, close the letter formally with "sincerely" or a similar polite expression. Sign your name and title. Be sure to provide correct, complete contact and reference information for future correspondence.
3. Do not forget to send the cover letter, with the RFP document, via certified mail.
4. Because things sometimes get a little more complicated than usual, remember to consult a lawyer for further information before doing anything.

[Location], [Date]

[Contact name]

[Address of the prospective provider]

Dear [Contact name]:

You are invited to submit a proposal for our [Project title] project in accordance with the requirements set forth in the attached *request for proposal* (RFP), which is also available online at [Site URL].

[Brief description of the company]

[Brief description of your existing systems and proposed project]

I will hold a [non-] mandatory preproposal meeting for prospective providers on [Date, start time] until [End time] at [Location]. Should you plan to attend the conference, you must [should] RSVP by [Date]. Questions must be received prior to the conference no later than [Date and time]. Both the RSVP and questions can be submitted to me in writing, by fax, or, preferably, by e-mail. Questions, answers, and modifications to the RFP will be posted on the Web at [Site URL] and will be debated publicly during the conference (see section [X] of the RFP).

If you intend to respond to the RFP, a letter of intent [which is not binding but will greatly assist me in planning for proposal evaluation] must [should] be submitted to me in writing, by fax or, preferably, by e-mail, and be received no later than [Date and time] (see section [X] of the RFP). [I will not accept your proposal if you do not complete the letter of intent within the specified time period.]

The original, [X] copies, and an electronic [Format] version of your proposal must be received no later than [Date and time] or your proposal will otherwise be disqualified. At the conference, you will receive preprinted labels to insure proper delivery and identification of your proposal (see section [X] of the RFP).

I anticipate that the provider whose proposal is the best solution for our project will be selected on [Date]. We will notify all providers, whether they are disqualified, rejected, or unsuccessful although responsive (see section [X] of the RFP).

I will be the single point of contact for all inquiries and correspondences.

I thank you for your time, effort, and interest in our [Project title] project.

Sincerely,

[Signature]

[Name and title of the person responsible for handling proposals]

The Request for Proposal (RFP) Letter of Intent

The RFP *letter of intent* tells the company issuing the RFP that you are interested not only in submitting a proposal in response, but also in receiving all RFP updates and modifications.

It is highly recommended to read the recommendations below in order to properly and successfully use the letter of intent.

1. Use a formal letterhead and do not handwrite the letter of intent.
2. First, *indicate your interest* in the RFP you received, and acknowledge the deadline for the proposal you will submit.
3. Next, remind the RFP issuer that you are, at the same time, *interested in being kept informed* about any modification related to this project (i.e., the RFP document).
4. Finally, *close the letter formally* with "sincerely" or a similar polite expression. Sign your name and title. Be sure to provide the correct, complete *contact and reference information* for future correspondence.
5. Do not forget to send the letter of intent via *certified mail.*
6. Because things sometimes get a little more complicated than usual, remember to consult a lawyer for further information before doing anything.

[Location], [Date]

[Name and title of the person responsible for handling proposals]

[Complete address]

Dear [Contact name]:

I would like to indicate our interest in the above Request for Proposal (RFP) and to be notified for any updates and amendments to the RFP.

Sincerely,

[Signature]

[Contact name]

[Complete address of the prospective provider]

[Phone and fax]

[E-mail address, an alias or distribution list dedicated to the RFP process]

The Letter to Decline a Proposal

The *letter to decline a proposal* is sent to a prospective provider in order to decline the proposal submitted in response to an RFP.

It is highly recommended to read the recommendations below in order to properly and successfully use the letter to decline the response submitted in response to the RFP.

1. Use a formal letterhead and do not handwrite an unsuccessful letter.
2. First, *thank the person* who submitted the proposal for the time, effort, and interest in the project related to the issued RFP.
3. Next, *notify the provider* that you decline the offer they have submitted because it is not the best apparent solution to the project for which the RFP was issued. The award is not officially definitive because unsuccessful providers have the right to formally contest their disqualification, rejection, or nonselection within a reasonable time frame, as initially defined in the RFP. Therefore, do not sign any contract with another provider until the deadline to receive protests expires and all protests are settled.
4. Your letter to decline the proposed response is not required to unveil information about to whom the project was awarded. Nevertheless, if requested, you must provide all information except for trade secrets.
5. Finally, *close the decline letter formally* with "sincerely" or a similar polite expression. Sign your name and title. Be sure to provide correct, complete contact and reference information for future correspondence.
6. Do not forget to send the decline letter via *certified mail.*
7. Because things sometimes get a little more complicated than usual, remember to consult a lawyer for further information before doing anything.

[Location], [Date]

[Contact name]

[Address of the company

that submitted the proposal

in response to the RFP]

Dear [Contact name]:

I thank you for your time, effort, and interest in our [Project Title] project.

Nevertheless, another proposal has been selected as the best solution relative to our project.

Sincerely,

[Signature]

[Name and title of the person responsible for handling disqualifications]

[Address]

The RFP Rejection Letter

The RFP *rejection letter* is sent to the prospective provider whose proposal has been rejected for specific reasons that are explicitly exposed in the rejection letter.

It is highly recommended to read the recommendations below in order to properly and successfully use the rejection letter.

1. Use a formal letterhead and do not handwrite a rejection letter.
2. First, thank the person who submitted the proposal for the time, effort, and interest in the project related to the issued RFP.
3. Next, state the reasons why the proposal is rejected. Be specific regarding these reasons. You must explain why and how the proposal is *nonresponsive* or the provider *nonresponsible*.
   Nonresponsiveness: A nonresponsive proposal would, for example, neglect to provide mandatory information or documents requested in the RFP.
   Nonresponsibility: A nonresponsible provider, although supplying all necessary information, would, for example, not be able to fully satisfy requirements defined in the RFP or would be financially unstable or unable to complete the project in a timely manner.
   Documenting the reasons why a proposal is rejected is far more difficult than merely identifying the proposal as noncompliant. Spend the time needed to honestly and properly communicate the reasons for the rejection. The more specific, exhaustive, and honest the reasons for the rejection are, the more difficult it becomes for the provider to contest your decision to reject the proposal.
4. Keep in mind that the rejected provider has the right to formally contest your decision within a reasonable time frame, as initially defined in the RFP. Therefore, do not sign any contract with another provider until the deadline to receive protests expires and rejection protests are settled.
5. You are not required to unveil information about the awarded company. Nevertheless, if requested, you must provide the information, except information labeled as trade secrets.
6. Finally, close the letter formally with "sincerely" or a similar polite expression. Sign your name and title.
7. Do not forget to send the rejection letter via *certified mail.*
8. Because things sometimes get a little more complicated than usual, remember to consult a lawyer for further information before doing anything.

[Location], [Date]

[Contact name]
[Address of the company
that submitted the proposal
in response to the RFP]

Dear [Contact name]:

I thank you for your time, effort, and interest in our [Project title] project.

Nevertheless, the proposal submitted has been rejected because of the following reasons:

[Reason 1]

[Detail]

...

[Reason X]

[Detail]

Sincerely,

[Signature]

[Name and title of the person responsible for handling rejections]

[Address]

The Sole Source Protest Letter

To properly and successfully file a protest against a procurement method or process based only on sole sourcing, you should write a sole source solicitation protest letter articulating your claim:

1. Use a *formal letterhead*; do not handwrite the protest letter.
2. First, *specify the action or decision being protested*, namely, the sole source contract. Describe how you, as a prospective provider, are affected by this action or decision.
3. Next, *prove* that the solicitation and selection process is clearly erroneous, arbitrary, or contrary to (a) the process defined in the solicitation document; (b) basic, generally accepted procurement policies, defined (for instance) in the Federal Acquisition Regulation (FAR) document setting forth basic policies and procedures for proper acquisition by all federal agencies; or even (c) law.

   Above all, the first valid (and thus acceptable) reason for your protest should be that your organization is *also* a capable, eligible, well-known, official, or competitive provider of the sole-sourced goods or services.

   Unless you give substance to your protest with a preponderance of evidence, your claim will be considered to be groundless, and will thus be dismissed.
4. Finally, *close the letter formally* (with "sincerely" or a similar polite expression). Sign your name and title. Be sure to provide complete *contact and reference information* for future correspondence.
5. Send the protest letter via *certified or registered mail.*
6. Because sometimes things are more complicated than usual, remember to consult a lawyer before doing anything.

[Date]

[Contact name]
[Address]

Dear [Contact name]:

As [your title within your organization], I am protesting the issuance by [Issuer's Name] of the [solicitation/contract] #[Solicitation/contract ID] to [Awarded Organization Name] as the sole source for [Project Title].

This protest is based on the following grounds:

[Evidence 1]

[Details and prejudice]

…

I hereby request a ruling by [Ruling entity name] on the actions to be performed as redress, as follows:

[Action 1]

[Details]

…

Looking forward to responding to a more competitive process, I thank you for your attention to this matter.

Sincerely,

[Your signature]

[Name, Title]

[Organization]

[Complete contact information]

The No-Bid Letter

A no-bid letter is a letter to the organization that invited you to bid or submit a proposal, notifying them that you will not do so. To remain potentially involved in future opportunities, the provider should state in the no-bid notice the reasons for declining such an invitation.

Before even writing a no-bid letter, you have to decide not to bid. This decision is the result of an analytical process, the bid/no-bid analysis.

Bid/No-Bid Analysis

The no-bid letter is sent after having performed a bid/no-bid analysis. The bid/no-bid analysis assesses (quantitatively, qualitatively, or both) all risks inherent in submitting or not submitting an offer. The analysis process relies on building a list of relevant questions, called the bid/no-bid checklist. On the basis of this checklist, a bid/no-bid analysis matrix will be created, which will determine the worth of sending a bid. If the decision is to bid, a letter of intent will be sent to the purchasing officer. If the decision is not to bid, then a no-bid letter, explaining the reasons, will be sent.

How to Stay on the Bidder List

The no-bid letter is the critical factor in remaining on the bidders list, and thus ensures future business opportunities. For the contracting officer who sent you the invitation to bid, the no-bid letter demonstrates that, although you are not interested in bidding for a particular project for specific and valid reasons, you are still interested in competing for future opportunities, and want to stay on the prospective bidder list. This is why it is important to take the time to write a professional no-bid letter.

It is highly recommended that you read the recommendations below in order use the no-bid letter template properly and successfully.

1. Use a formal letterhead. Do not handwrite your no-bid letter.
2. First, your no-bid letter should *thank the person* who sent you the invitation, for showing interest in the solution your organization is marketing.
3. Next, *state the reasons why you are not proposing an offer or bidding.* Be specific regarding these reasons. The best way to discover valid, thus acceptable reasons not to bid is to perform a *bid/no-bid analysis.* This step is the critical factor in remaining on the list of prospective providers.
4. Reiterate that you are still *interested in being kept informed* about new business opportunities, for which your solution may be well suited, or best suited.

5. Finally, *end the letter formally* (with "sincerely," for example, or a similar polite expression). Sign your name and indicate your title. Be sure to provide correct and complete *contact and reference information* for future correspondence.
6. Before sending it, make sure your no-bid letter is tactful, respectful, and gets straight to the point.
7. Send the no-bid letter via *registered mail*. It has to be received before the bid/proposal opening date.
8. Because things sometimes are a little more complicated, remember to consult a lawyer for information before doing anything.

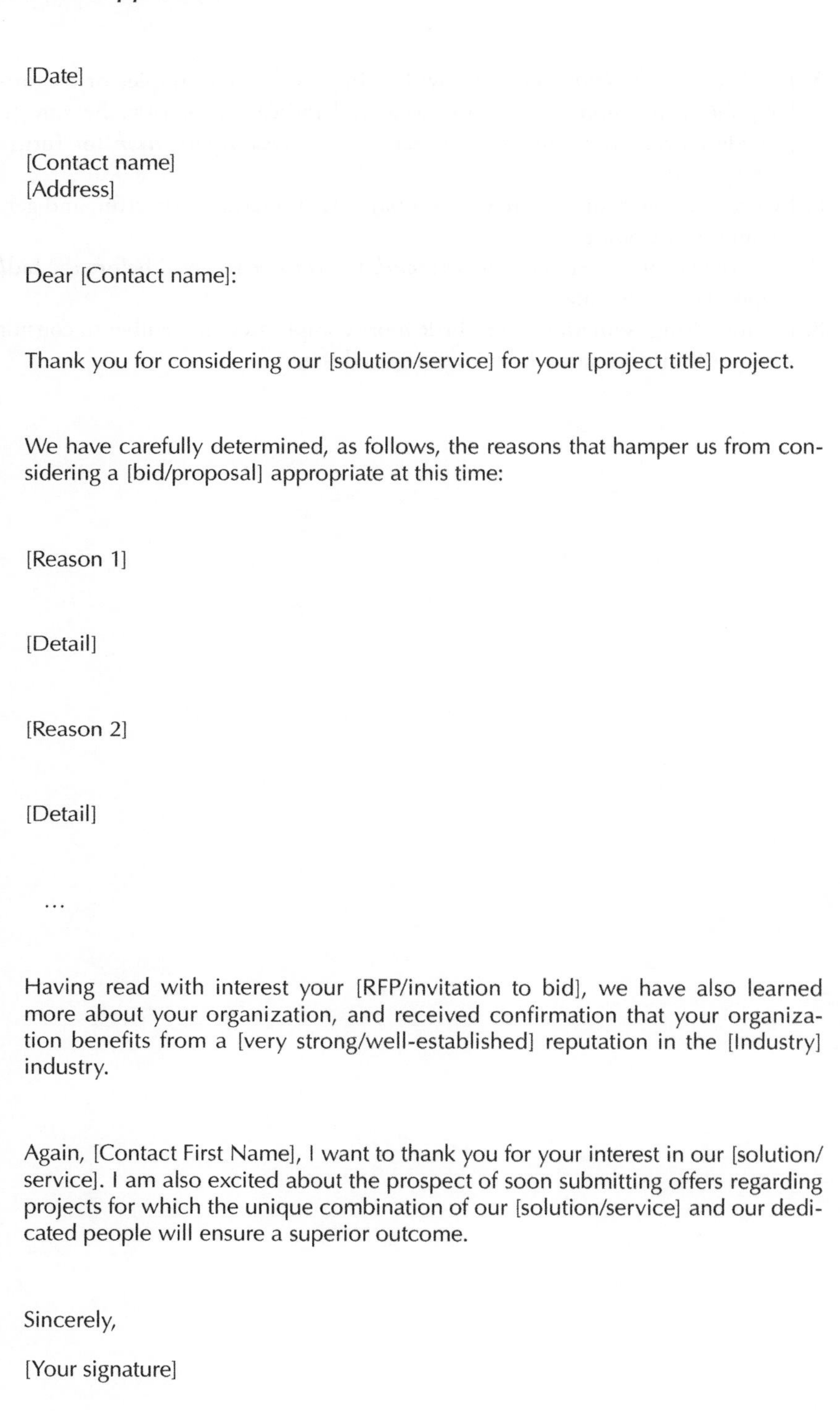

[Date]

[Contact name]
[Address]

Dear [Contact name]:

Thank you for considering our [solution/service] for your [project title] project.

We have carefully determined, as follows, the reasons that hamper us from considering a [bid/proposal] appropriate at this time:

[Reason 1]

[Detail]

[Reason 2]

[Detail]

...

Having read with interest your [RFP/invitation to bid], we have also learned more about your organization, and received confirmation that your organization benefits from a [very strong/well-established] reputation in the [Industry] industry.

Again, [Contact First Name], I want to thank you for your interest in our [solution/service]. I am also excited about the prospect of soon submitting offers regarding projects for which the unique combination of our [solution/service] and our dedicated people will ensure a superior outcome.

Sincerely,

[Your signature]

[Name, Title]

[Organization]

[Exhaustive contact information]

The RFP Disqualification Letter

The RFP *disqualification letter* is sent to the prospective provider whose proposal has been disqualified for specific reasons that are explicitly exposed in the disqualification letter.

It is highly recommended to read the recommendations below in order to properly and successfully use the disqualification letter.

1. Use a formal letterhead and do not handwrite a disqualification letter.
2. First, thank the person who submitted the proposal for the time, effort, and interest in the project related to the issued RFP.
3. Next, explain the reasons why and how the proposal is disqualified. Specify the date when disqualification is effective. Be specific regarding these reasons. If you want the disqualification argument to be as valid as possible, it is highly recommended not to open the envelope containing the proposal. Return the proposal, as is, whether you opened it or not, accompanied by the disqualification letter.

   Document the reasons why the proposal is disqualified and spend the time needed to honestly and properly communicate the reasons for the disqualification. The more specific, exhaustive, and honest the reasons for the disqualification are, the more difficult it becomes for the provider to contest your decision to disqualify the proposal.
4. Keep in mind that the disqualified provider has the right to formally contest your decision within a reasonable time frame, as initially defined in the RFP. Therefore, do not sign any contract with another provider until the deadline to receive protests expires and disqualification protests are settled.
5. You are not required to unveil information about the awarded company. Nevertheless, if requested, you must provide the information, except that labeled as trade secrets.
6. Finally, close the letter formally with "sincerely" or a similar polite expression. Sign your name and title.
7. Do not forget to send the disqualification letter via certified mail.
8. Because things sometimes get a little more complicated than usual, remember to consult a lawyer for further information before doing anything.

[Location], [Date]

[Contact name]
[Address of the company
that submitted the proposal
in response to the RFP]

Dear [Contact name]:

I thank you for your time, effort, and interest in our [Project Title] project. Nevertheless, you will find enclosed and unopened the proposal you have submitted and that has been disqualified for the following reasons:

1. [Reason 1]

[Detail]

...

X. [Reason X]

[Detail]

You may appeal this disqualification by notifying us in writing within [X] (X) business days after receipt of this notification.

Sincerely,

[Signature]

[Name and title of the person responsible for handling disqualifications]

[Address]

The RFP Protest Letter

The RFP *Protest Letter* may be sent by a prospective provider who is aggrieved in connection with the RFP specifications, the solicitation process, or award of the contract and would like to file a protest. The scope of the grounds of the protest is limited to errors related to proper proposal scoring, violation of law or procedures set forth in the RFP, conflict of interest, partiality, or discrimination. No protest may be filed if the RFP is cancelled or if all proposals received in response to the RFP are rejected. The RFP protest letter is presented on the next page. It is highly recommended to read the recommendations below in order to properly and successfully use the RFP protest letter.

1. Use a formal letterhead and do not handwrite the protest letter.
2. First, *specify the precise action or decision being protested*, and describe how you, as a prospective provider, have been or would be affected by or aggrieved with this action or decision.
3. Next, *you have to prove* that the evaluation, comparison, and selection process is clearly erroneous, arbitrary, or contrary to the process defined in the RFP or even to law. As long as the person responsible for evaluating the proposals follows the requirements set forth in the RFP, all proposals are considered fairly and in good faith.
   Solely stating that your proposal is better than others is a fruitless claim. Faulting to give substance to your protest by a preponderance of evidence, your claim will be considered as a mere groundless allegation, and will be thusly dismissed.
4. Finally, *close the letter formally* with "sincerely" or a similar polite expression. Sign your name and title. Be sure to provide the correct, complete *contact and reference information* for future correspondence.
5. Do not forget to send the protest letter via *certified mail*.
6. Because things sometimes get a little more complicated than usual, remember to consult a lawyer for further information before doing anything.

[Location], [Date]

[Name and title of the person responsible for handling proposals]
[Complete address]

Dear [Contact name]:

On [Date], we have submitted a proposal in accordance to the requirements set forth in the Request for Proposal for your [Project title] project.

On [Date], we received a letter indicating that our proposal was not selected [disqualified or rejected] for the following reasons:

1. [Reason 1]

...

X. [Reason X]

The purpose of this letter, in accordance to the conditions set forth in section [X] of the RFP, is to protest the nonselection [disqualification or rejection] of our proposal by providing evidence that:

1. [Evidence 1]

[Detail]

...

X. [Evidence X]

[Detail]

Sincerely,

[Signature]

[Contact name]

[Complete address of the prospective provider]

[Phone and fax]

[E-mail address, an alias or distribution list dedicated to the RFP process]

The Contract Award Letter

The *contract award letter* is sent to the provider with the solution that, for the best value, fully satisfies or best addresses the requirements defined in the RFP.

It is highly recommended to read the recommendations below in order to properly and successfully use the contract award letter.

1. Use a formal letterhead and do not handwrite a contract award letter.
2. First, thank the person who submitted the proposal for the time, effort, and interest in the project related to the issued RFP.
3. Next, notify the provider that the submitted proposal is the best apparent solution to the project for which the RFP was issued. The award is not officially definitive, because unsuccessful providers have the right to formally contest their disqualification, rejection, or nonselection within a reasonable time frame, as initially defined in the RFP. Therefore, do not sign any contract with the selected provider until the deadline to receive protests expires and all protests are settled.
4. Finally, close the letter formally with "sincerely" or a similar polite expression. Sign your name and title.
5. Do not forget to send the contract award letter via *certified mail.*
6. Because things sometimes get a little more complicated than usual, remember to consult a lawyer for further information before doing anything.

[Location], [Date]

[Contact name]
[Address of the company
that submitted the proposal
in response to the RFP]

Dear [Contact name]:

I thank you for your time, effort, and interest in our [Project title] project.

Furthermore, I am pleased to announce that your proposal is the best apparent successful solution relative to our project.

Upon termination of the required protest period of [X] (X) business days after the closing date of [Closing date and time], I will be contacting you in order to execute a formal Professional Service Agreement (PSA).

Sincerely,

[Signature]

[Name and title of the person responsible for handling disqualifications]

[Address]

U.S. Geological Survey Manual

402.5—Procurement Specifications

8/7/95

OPR: Admin/Office of Procurement and Contracts

1. Purpose. This chapter provides guidelines and procedures for developing specifications and statements of work, including the use of Federal specifications and standards, used in the acquisition of supply items, data and services.
2. Authority. Federal Acquisition Regulation (FAR) 10.006 requires all Federal agencies to use specifications and standards listed in the Index of Federal Specifications and Standards. Deviations from the use of prescribed specifications are described in FAR 10.007.

   The Metric Conversion Act of 1975 and the Omnibus Trade and Competitiveness Act of 1988, as implemented in 758 DM 1, require use of metric measurements in specifications whenever practicable.

   The Resource Conservation and Recovery Act (RCRA) requires government agencies to buy products containing the maximum feasible amount of recovered and/or recycled materials. Together, RCRA; the Pollution Prevention Act of 1990; Executive Order 12759, Federal Energy Management; and the Comprehensive Environmental Response, Compensation and Liability Act, make it a national policy to (1) prevent pollution by purchasing and using products and services which are environmentally sound and energy efficient, and (2) reduce pollution by buying and using products containing recovered or recycled materials. The Environmental Protection Agency (EPA) has identified various types of items referred to as the "EPA Guideline Items" which require either recycled/recovered material content, or a waiver by the Head of the Procuring Activity based on one or more of the following exceptions:

   A. The item with recycled/recovered content does not meet all reasonable performance specifications. Use of such an item would jeopardize the intended end use of the designated item.
   B. An item containing recycled/recovered materials is not readily available at a reasonable price.
   C. Applying minimum content standards results in inadequate competition.

D. Use of an item containing recycled/recovered materials would pose a safety problem.

E. Obtaining items with recycled/recovered material content would pose an unusual and unreasonable delay.

Executive Order 12843 requires agencies to conform their procurement practices to the requirements of the Clean Air Act Amendments (Public Law 101-549) to reduce use of substances that cause stratospheric ozone depletion.

3. Definitions.

A. Specification means a physical and/or functional description of a material, product, data or service needed to fill a USGS requirement, including the criteria for determining whether or not the requirements are met.

B. Standard means a document that establishes engineering and technical limitations and applications of items, materials, processes, methods, designs, and engineering practices. It includes any related criteria deemed essential to achieve the highest degree of uniformity in materials or products, or interchangeability of parts used in those products. Standards may be used in specifications, solicitations, and contracts.

C. Federal specification or standard means a specification or standard issued or controlled by GSA and listed in the Index of Federal Specifications, Standards and Commercial Item Descriptions. Federal specifications must be used unless (1) the value of the purchase is below the small purchase threshold; (2) the item is a commercial item, for which you may use a purchase description or commercial item description; or, (3) no federal specification exists for the item(s) being procured.

D. Voluntary standard, also known as nongovernment standard, means a standard established by a private sector association, organization or technical society, and available for public use. The term does not include private standards of individual firms. For further guidance, see OMB Circular A-119, Federal Participation in Development and Use of Voluntary Standards.

E. Commercial item description means an indexed, simplified product description managed by the General Services Administration (GSA), which describes, by functional or performance characteristics, the available, acceptable commercial products capable of satisfying the Government's needs.

F. Product description is the generic term for documents used for acquisition and management purposes, such as specifications, standards, voluntary standards, commercial item descriptions, or purchase descriptions.

G. Purchase description means any product description containing the essential physical characteristics and functions required to meet the Government's minimum needs, usually prepared for one-time use, for small purchases, or when development of an indexed product description is not otherwise cost effective.

H. Metric measurements are statements of size, volume, or weight units based on a decimal system which uses the meter, liter, and the kilogram in lieu of the inch, gallon, and pound, as is traditional in the English system. It is used as the standard measurement system throughout most of the world.

I. Dual system is a method of stating measurement units in a specification in such a way that both the Metric and English measurement units are available for a vendor's use. A dual system of measurements is needed during the period of transition to metric measurements in the United States to insure that we do not eliminate non-metric vendors from competition until such time as they can convert their manufacturing capabilities to comply with standard metric measurements.

J. EPA Guideline Items are the specific categories of items identified by the Environmental Protection Agency (EPA) which must be procured in forms containing the maximum amount of post-consumer, recovered/recycled materials practicable, economic and performance factors considered. EPA's Comprehensive Procurement Guideline for Products Containing Recovered Materials (Title 40, Code of Federal Regulations, part 247) is provided in Appendix A. Specific recommendation for recycled content for each of the Guideline items is contained in the EPA Recovered Materials Advisory Notice, Section II (Appendix B).

4. Responsibilities.

A. The Head of the Contracting Activity (Assistant Director for Administration) is responsible for authorizing deviations from use of Federal specifications under FAR 10.007 and Department of the Interior Acquisition Regulation (DIAR) 1410.007, and for granting exceptions to the use of Environmental Protection Agency (EPA) guidelines for items containing recovered or recycled products.

B. Division Chiefs are responsible for monitoring the requisitions created in their divisions to insure that program officials actively pursue the use of products that enable the USGS to eliminate, reduce, recycle or better manage its solid wastes and/or hazardous wastes and ozone-depleting substances. Division Chiefs are also responsible for determining when program requirements preclude acquisition of products with recycled content, except those items covered by EPA guidelines, and for determining which system of measures is appropriate for each specification under their control.

C. Requisitioners and Requisition Reviewing Officials are responsible for preparing their requests for supplies and services within the framework of and with a full knowledge of specifications and standards applicable to their cognizant areas of program responsibility. They are responsible for the requisition of commercial items whenever they are available and suitable for our needs, using standard Government specifications whenever possible, and for drafting specifications and statements of work which:
   (1) Are accurate and clear;
   (2) Are complete and concise, reflecting only USGS minimum needs;
   (3) Promote competition;
   (4) Are quantifiable;
   (5) Do not contain unnecessarily restrictive requirements;
   (6) Encourage the acquisition of commercially available items;
   (7) Eliminate or reduce the use of hazardous materials and the creation of hazardous wastes;
   (8) Require recycled or recovered materials to the maximum extent practical and safe, stressing the purchase/use of EPA guideline items,
   (9) Eliminate or reduce use of ozone-depleting substances by substitution of alternate substances or new technology to the extent economically practical;
   (10) Ensure the acquisition and use of goods and services which are otherwise environmentally sound and energy efficient, and
   (11) Use metric measurements or a dual (metric/inch-pound) system of dimensions, except when metric use is impractical or is likely to cause inefficiencies or loss of markets to United States firms.

   The specification review sheet provided as Figure 1 documents the requiring office's considerations regarding specification of recycled content and metric conversion requirements. This form is to be completed and provided to the procuring office with requisitions using a specification or statement of work having substantial design elements.

D. Contracting Officers are responsible for ensuring that the appropriate types of specifications and standards are used in all solicitations and resultant contracts entered into by the USGS for the procurement of supplies and services to fill official requirements.

5. Types of Specifications.

There are several types of specifications. One of these should accompany any DI1 requisition form submitted to the contracting office for processing. The different types of specifications are:

A. Performance specifications describe the product/service in terms of form, fit and function. "Form" deals with the general constraints to be placed on the item/service. "Fit" describes how the procured item or service will be compatible with related or existing item/service. "Function" describes what the article must do. Performance specifications are the preferred type of specifications for USGS use.

B. Design specifications describe an item in terms of its detailed form or composition, i.e., specific materials, dimensions, design concepts, methods of manufacture, etc. This type of specification stipulates a design from which a supplier cannot depart in order to substitute its own design preference. When manufacturing or service problems arise from faulty or unclear specifications, the Government, as drafter of the ambiguous specification, is almost always found to be liable for any remedies required.

C. Hybrid specifications combine the design and performance specifications.

D. Formulation (Material) specifications define in detail the chemical composition of the material. In addition, it may define the manufacturing or production techniques to be used to blend the chemicals into the desired end product. This type of specification is used mainly for the purchase of metals, chemicals, plastics, paints, etc.

E. Brand Name or Equal Descriptions identify products by their brand name, make, model or catalog number and name and address of the manufacturer. (See FAR 10.001 and DIAR 1410.00470). Offerors can propose an equal product that has the same salient characteristics as the brand name product.

F. Specific Make and Model specifications (also called "Brand Name Descriptions") identify the specific make and the model number of the item to be procured. (See FIRMR 20139.601

and 20120.103-5.) As specified in FIRMR 20139.601-3, this type specification should only be used when no other type of specification can satisfy the Government's needs. The make and model specification differs from the brand name or equal in that potential contractors cannot propose an equal item. A specific make and model specification restricts competition and must be accompanied by a justification for other than full and open competition in accordance with FAR 6.303.

6. The Statement of Work.

Each statement of work (SOW) describes (1) the service to be provided by the contractor, (2) the conditions under which the work must be conducted, (3) an explanation of how the contractor's achievement will be assessed, (4) reports, data, and items to be delivered by the contractor, and (5) any other contractor and government obligations related to the procurement. The SOW must be clear, complete, and precise in describing the supplies and or services in a manner designed to promote full and open competition as well as compliance with applicable laws and regulations.

An SOW which is clearly defined and understood by the both the government and the potential contractor will ensure that the government (1) receives the item, data, or service necessary to fill its needs; (2) pays a price which is fair and reasonable to both the government and the contractor; and (3) spends a minimum effort on contract administration. A well defined SOW saves time and effort throughout the acquisition process. Ambiguities in Government prepared specifications result in costly delays and additional effort by the contractor, for which the Government, as drafter of the specifications, must bear the expense.

It is always preferable to develop and use a performance oriented, or functional, type SOW. The SOW should identify "what" is to be delivered, or the end result of the service, not "how" the work is to be accomplished. The tasks should be expressed in clear, concise, and enforceable terms.

The development and use of a performance oriented SOW, rather than a detailed specification, results in a contract which places the maximum risk of performance on the contractor. Detailed specifications, in which the Government describes exactly how the item is to be created or the service is to be performed, result in the maximum risk of performance being held against the Government. If detailed specifications are used in lieu of performance specifications, the contractor can only be required to meet the requirements contained in the detailed specifications, whether or not the desired performance criteria are met. SOWs for follow-on procurements should be reviewed and revised to remove any ambiguities and make them more performance oriented.

U.S.C. Section 3729—False Claims

(a) *Liability for Certain Acts.*—Any person who—

(1) knowingly presents, or causes to be presented, to an officer or employee of the United States Government or a member of the Armed Forces of the United States a false or fraudulent claim for payment or approval;

(2) knowingly makes, uses, or causes to be made or used, a false record or statement to get a false or fraudulent claim paid or approved by the Government;

(3) conspires to defraud the Government by getting a false or fraudulent claim allowed or paid;

(4) has possession, custody, or control of property or money used, or to be used, by the Government and, intending to defraud the Government or willfully to conceal the property, delivers, or causes to be delivered, less property than the amount for which the person receives a certificate or receipt;

(5) authorized to make or deliver a document certifying receipt of property used, or to be used, by the Government and, intending to defraud the Government, makes or delivers the receipt without completely knowing that the information on the receipt is true;

(6) knowingly buys, or receives as a pledge of an obligation or debt, public property from an officer or employee of the Government, or a member of the Armed Forces, who lawfully may not sell or pledge the property; or

(7) knowingly makes, uses, or causes to be made or used, a false record or statement to conceal, avoid, or decrease an obligation to pay or transmit money or property to the Government,

is liable to the United States Government for a civil penalty of not less than $5,000 and not more than $10,000, plus 3 times the amount of damages which the Government sustains because of the act of that person, except that if the court finds that—

(A) the person committing the violation of this subsection furnished officials of the United States responsible for investigating false claims violations with all information known to such person about the violation within 30 days after the date on which the defendant first obtained the information;

(B) such person fully cooperated with any Government investigation of such violation; and

(C) at the time such person furnished the United States with the information about the violation, no criminal prosecution, civil action, or administrative action had commenced under this title with respect to such violation, and the person did not have actual knowledge of the existence of an investigation into such violation;

the court may assess not less than 2 times the amount of damages which the Government sustains because of the act of the person. A person violating this subsection shall also be liable to the United States Government for the costs of a civil action brought to recover any such penalty or damages.

(b) *Knowing and Knowingly Defined.*— For purposes of this section, the terms "knowing" and "knowingly" mean that a person, with respect to information—
  (1) has actual knowledge of the information;
  (2) acts in deliberate ignorance of the truth or falsity of the information; or
  (3) acts in reckless disregard of the truth or falsity of the information,

  and no proof of specific intent to defraud is required.

(c) *Claim Defined.*— For purposes of this section, "claim" includes any request or demand, whether under a contract or otherwise, for money or property which is made to a contractor, grantee, or other recipient if the United States Government provides any portion of the money or property which is requested or demanded, or if the Government will reimburse such contractor, grantee, or other recipient for any portion of the money or property which is requested or demanded.

(d) *Exemption from Disclosure.*— Any information furnished pursuant to subparagraphs (A) through (C) of subsection (a) shall be exempt from disclosure under section 552 of title 5.

(e) *Exclusion.*— This section does not apply to claims, records, or statements made under the Internal Revenue Code of 1986.

15 U.S.C. Section 13—Discrimination in Price, Services, or Facilities

(a) *Price; selection of customers.* It shall be unlawful for any person engaged in commerce, in the course of such commerce, either directly or indirectly, to discriminate in price between different purchasers of commodities of like grade and quality, where either or any of the purchases involved in such discrimination are in commerce, where such commodities are sold for use, consumption, or resale within the United States or any Territory thereof or the District of Columbia or any insular possession or other place under the jurisdiction of the United States, and where the effect of such discrimination may be substantially to lessen competition or tend to create a monopoly in any line of commerce, or to injure, destroy, or prevent competition with any person who either grants or knowingly receives the benefit of such discrimination, or with customers of either of them: Provided, That nothing herein contained shall prevent differentials which make only due allowance for differences in the cost of manufacture, sale, or delivery resulting from the differing methods or quantities in which such commodities are to such purchasers sold or delivered: Provided, however, That the Federal Trade Commission may, after due investigation and hearing to all interested parties, fix and establish quantity limits, and revise the same as it finds necessary, as to particular commodities or classes of commodities, where it finds that available purchasers in greater quantities are so few as to render differentials on account thereof unjustly discriminatory or promotive of monopoly in any line of commerce; and the foregoing shall then not be construed to permit differentials based on differences in quantities greater than those so fixed and established: And provided further, That nothing herein contained shall prevent persons engaged in selling goods, wares, or merchandise in commerce from selecting their own customers in bona fide transactions and not in restraint of trade: And provided further, That nothing herein contained shall prevent price changes from time to time where in response to changing conditions affecting the market for or the marketability of the goods concerned, such as but not limited to actual or imminent deterioration of perishable goods, obsolescence of seasonal goods, distress sales under court process, or sales in good faith in discontinuance of business in the goods concerned.

15 U.S.C. Section 13—Discrimination in Price, Services, or Facilities

(b) *Burden of rebutting prima-facie case of discrimination.* Upon proof being made, at any hearing on a complaint under this section, that there has been discrimination in price or services or facilities furnished, the burden of rebutting the prima-facie case thus made by showing justification shall be upon the person charged with a violation of this section, and unless justification, and unless justification shall be affirmatively shown, the Commission is authorized to issue an order terminating the discrimination: Provided, however, That nothing herein contained shall prevent a seller rebutting the prima-facie case thus made by showing that his lower price or the furnishing of services or facilities to any purchaser or purchasers was made in good faith to meet an equally low price of a competitor, or the services or facilities furnished by a competitor.

15 U.S.C. Section 13—Discrimination in Price, Services, or Facilities

(c) *Payment or acceptance of commission, brokerage or other compensation.* It shall be unlawful for any person engaged in commerce, in the course of such commerce, to pay or grant, or to receive or accept, anything of value as a commission, brokerage, or other compensation, or any allowance or discount in lieu thereof, except for services rendered in connection with the sale or purchase of goods, wares, or merchandise, either to the other party to such transaction or to an agent, representative, or other intermediary therein where such intermediary is acting in fact for or in behalf, or is subject to the direct or indirect control, of any party to such transaction other than the person by whom such compensation is so granted or paid.

15 U.S.C. Section 13—Discrimination in Price, Services, or Facilities

(d) *Payment for services or facilities for processing or sale.* It shall be unlawful for any person engaged in commerce to pay or contract for the payment of anything of value to or for the benefit of a customer of such person in the course of such commerce as compensation or in consideration for any services or facilities furnished by or through such customer in connection with the processing, handling, sale, or offering for sale of any products or commodities manufactured, sold, or offered for sale by such person, unless such payment or consideration is available on proportionally equal terms to all other customers competing in the distribution of such products or commodities.

15 U.S.C. Section 13—Discrimination in Price, Services, or Facilities

(e) *Furnishing services or facilities for processing, handling, etc.* It shall be unlawful for any person to discriminate in favor of one purchaser against another purchaser or purchasers of a commodity bought for resale, with or without processing, by contracting to furnish or furnishing, or by contributing to the furnishing of, any services or facilities connected with the processing, handling, sale, or offering for sale of such commodity so purchased upon terms not accorded to all purchasers on proportionally equal terms.

15 U.S.C. Section 13—Discrimination in Price, Services, or Facilities

(f) Knowingly inducing or receiving discriminatory price. It shall be unlawful for any person engaged in commerce, in the course of such commerce, knowingly to induce or receive a discrimination in price which is prohibited by this section. Enacted: October 15, 1914, as amended June 19, 1936. Section 3 of the Robinson-Patman Act, 15 U.S.C.A. Section 13a—Not an "Antitrust" Statute

Section 3 of the Robinson-Patman Act, 15 U.S.C.A. Section 13a—Not an "Antitrust" Statute

Note: Before looking at Section 3 of the Robinson-Patman Act, you should bear in mind that the statute is only a penal statute and therefore has been held not to be an "antitrust" statute for purposes of the treble-damage and other private remedies available under the Sherman Act and other provisions of the Robinson-Patman Act.

The following is quoted from *Englander Motors, Inc. v. Ford Motor Company*, 293 F.2d 802, *; 1961 U.S. App. LEXIS 3903, **; 1961 Trade Cas. (CCH) P70,070 (6th Cir. 1961), provides the explanation and citation to the relevant U.S. Supreme Court decision:

The action in the District Court was to recover treble damages under Section 4 of the Clayton Act for alleged violations of Sections 2(a) and 3 of the Robinson-Patman Act. 15 U.S.C.A. §§ 15, 13, 13a.

Defendant filed a motion for summary judgment on two grounds, namely, (1) that an action for treble damages under Section 4 of the Clayton Act cannot be brought for violations of Section 3 of the Robinson-Patman Act and (2) the action is barred by the Ohio statute of limitations. The District Court Granted the motion on both grounds and dismissed the complaint. Plaintiff appealed.

In our judgment, the District Court was correct in holding that the action for treble damages under Section 4 of the Clayton Act could not be maintained for violations of Section 3 of the Robinson-Patman Act. The latter section is not an "antitrust law" within the meaning of Section 4 of the Clayton Act. Section 3 of the Robinson-Patman Act contains only penal sanctions for its violation. These are exclusive. The private remedy afforded by Section 4 of the Clayton Act cannot, therefore, be based on violations of Section 3 of the Robinson-Patman Act. *Nashville Milk Co. v. Carnation Co.*, 1958, 355 U.S. 373, 78 S.Ct. 352, 2 L.Ed.2d 340; *Safeway Stores, Inc. v. Vance*, 1958, 355 U.S. 389, 78 S.Ct. 358, 2 L.Ed.2d 350; *Ludwig v. American Greetings Corp.*, 6 Cir., 1959, 264 F.2d 286.

§ 13a. Discrimination in rebates, discounts, or advertising service charges; underselling in particular localities; penalties

It shall be unlawful for any person engaged in commerce, in the course of such commerce, to be a party to, or assist in, any transaction of sale, or contract to sell, which discriminates to his knowledge against competitors of the purchaser, in that, any discount, rebate, allowance, or advertising service charge is granted to the purchaser over and above any discount, rebate, allowance, or advertising service charge available at the time of such transaction to said competitors in respect of a sale of goods of like grade, quality, and quantity; to sell, or contract to sell, goods in any part of the United States at prices lower than those exacted by said person elsewhere in the United States for the purpose of destroying competition, or eliminating a competitor in such part of the United States; or, to sell, or contract to sell, goods at unreasonably low prices for the purpose of destroying competition or eliminating a competitor.

Any person violating any of the provisions of this section shall, upon conviction thereof, be fined not more than $ 5000 or imprisoned not more than one year, or both.

[History: Statute enacted on 6/19/36, ch 592, § 3, 49 Stat. 1528, as part of the Robinson-Patman Act]

15 U.S.C. Section 26

Section 26. Injunctive relief for private parties; exception; costs

Any person, firm, corporation, or association shall be entitled to sue for and have injunctive relief, in any court of the United States having jurisdiction over the parties, against threatened loss or damage by a violation of the antitrust laws, including sections two, three, seven and eight of this Act, when and under the same conditions and principles as injunctive relief against threatened conduct that will cause loss or damage is granted by courts of equity, under the rules governing such proceedings, and upon the execution of proper bond against damages for an injunction improvidently granted and a showing that the danger of irreparable loss or damage is immediate, a preliminary injunction may issue: Provided, That nothing herein contained shall be construed to entitle any person, firm, corporation, or association, except the United States, to bring suit for injunctive relief against any common carrier subject to the jurisdiction of the Surface Transportation Board under subtitle IV of title 49, United States Code. In any action under this section in which the plaintiff substantially prevails, the court shall award the cost of suit, including a reasonable attorney's fee, to such plaintiff.

Antitrust Statute

Last Update: 8/3/01 18:00

Antitrust Statute: Section 1 of the Sherman Antitrust Act, 15 U.S.C. Section 1

Section 1. Trusts, etc., in restraint of trade illegal; penalty

Every contract, combination in the form of trust or otherwise, or conspiracy, in restraint of trade or commerce among the several states, or with foreign nations, is declared to be illegal. Every person who shall make any contract or engage in any combination or conspiracy hereby declared to be illegal shall be deemed guilty of a felony, and, on conviction thereof, shall be punished by fine not exceeding $10,000,000 if a corporation, or, if any other person, $350,000, or by imprisonment not exceeding three years, or by both said punishments, in the discretion of the court.

[As of 12/21/74]

Antitrust Statute
Last Update: 8/3/01 18:00

Antitrust Statute: Section 2 of the Sherman Antitrust Act, 15 U.S.C. Section 2

Section 2. Monopolizing trade a misdemeanor; penalty

Every person who shall monopolize, or attempt to monopolize, or combine or conspire with any other person or persons, to monopolize any part of the trade or commerce among the several States, or with foreign nations, shall be deemed guilty of a felony, and, on conviction thereof, shall be punished by fine not exceeding $10,000,000 if a corporation, or, if any other person, $350,000, or by imprisonment not exceeding three years, or by both said punishments, in the discretion of the court.

[As amended 12/21/74]

Ethical Procurement

Ethics dictates that business be conducted with integrity, fairness, and openness, which require open communication within and among both buying and supplying organizations, ensuring a competitive procurement, and thereby give any organization a chance to compete for contracts and win. It seems like a week does not go by without one hearing a story demonstrating lapses in business ethics stories, where lucky contract winners share close friends in strategic places, as was the case of *Halliburton*, which was awarded billions in federal no-bid contracts during the years when *Dick Cheney* was its chief executive officer (CEO).

Recognizing this longstanding problem of competitive procurement becoming the exception, the U.S. government launched the *Federal Acquisition Regulation* (FAR) more than 20 years ago to help federal agencies manage procurement more efficiently. FAR is an official document setting forth procurement policies and procedures that U.S. federal agencies should follow when soliciting offers (bids or proposals) for goods, products, services, or construction from qualified suppliers. Rules are delineated into eight chapters:

1. General Information, Definitions, Practices, and Administrative Matters
2. Competition and Acquisition Planning
3. Contracting Methods and Contract Types
4. Socioeconomic Programs
5. General Contracting Requirements
6. Special Categories of Contracting
7. Contract Management
8. Clauses and Forms

In response to competition becoming the exception rather than the rule in procurement, Part 6 of the FAR, *Competitive Requirements*, incorporated the *Competition in Contracting Act* (CICA). It seeks to establish *full and open competition* (FOC) as the standard, in order to eradicate partiality, favoritism, political lobbying, and bribery.

In light of these corporate scandals and the heightened scrutiny over corporate practices, FAR will have a greater impact on business. It contains key elements that all organizations should be aware of, whether they are conducting business with a government agency or with private entities. Using the policies and procedures outlined by FAR, we will

- Define what exactly is a *competitive procurement*, its purpose and benefits
- Present the different *competitive* and *noncompetitive* procurement methods and their supporting processes

Detail the two key procurement methods:

1. Invitation for bids (IFB)
2. Request for proposals (RFP)

Competitive Procurement

Also known as *Full and Open Competition (FOC)*, *competition-based procurement*, *competitive procurement process*, and *competitive solicitation*. competition-based purchase process, competitive contractual acquisition, Competition In Contracting Act, CICA, competitive acquisition, competitive procurement processes, competitive purchase process, competition-based acquisition, federal competitive procurement, competitive procurement methods, Request for Proposals, RFP, RFP procurement, RFP solicitation, Invitation for Bids, IFB procurement, IFB solicitation, Invitation to Bid, ITB, ITB procurement, ITB solicitation, competitive procurement policies and procedures, competitive procurement methods, Federal Acquisition Regulation, or FAR.

Benefits of Competitive Procurement

The purpose of competition is to benefit your enterprise, and competition should not be promoted for the sake of competition. In some cases, it may cost more to enter into a competitive procurement to do business directly with a supplier because of the *unicity* of the requested products, the *inadequacy* of other sources, the *immediacy* of your needs, or the *emergency* or *legitimacy* of circumstances. In other words, competitive procurement should be promoted pragmatism, with other methods thus becoming exceptions.

Competitive procurement is the contractual acquisition (purchase or lease) by an organization of any kind of asset, whether material (goods, products, or construction) or immaterial (services) with appropriated funds, enabling all deemed responsible sources to compete in a fair and open environment. Competitive procurement is also known as *full and open competition* (FOC), or *competitive solicitation*.

Although procurement is supposed to bring fairness, impartiality, transparency, and suitability to corporate practices, a competitive procurement process will ensure the highest level of openness, thus maximizing the suitability of the requested assets or services, and the best return on investment (ROI).

Different Procurement Methods

There are several procurement methods that can be used to complete an acquisition. Let us define the difference between procurement methods and their supporting processes. A procurement method is the manner chosen to perform a contractual acquisition. Procurement methods could be any of the three following:

1. Simplified acquisitions, like micropurchases, small purchases
2. Full and open competition (FOC), like bids, proposals
3. Other than full and open competition (OFOC), like sole source.

Simplified acquisitions procedures are the procurement methods used for acquisitions for which the amount does not exceed $2500 for the micropurchase threshold, and $100,000 for the small purchase threshold (FAR 2.101). Simplified acquisitions are equitably distributed among qualified suppliers in the local area, and purchases should not be split to avoid the requirements for competition above their respective thresholds. Minimal documentation is usually required, including a determination that the price is fair and reasonable; and material on how this determination was derived.

Full and Open Competition (FOC)

When do you have to put a formal solicitation process based on competition in place? You may be required, depending on your organization's purchasing policy, to use a formal solicitation and selection (i.e., competitive procurement, process for acquisitions for which the amount exceeds a certain threshold—in our case, the simplified acquisition threshold).

The competitive procurement methods available for use in fulfilling the requirement for full and open competition in the acquisition process are

1. *Sealed bidding*, relying on *bids*, also called *sealed bids*
2. *Negotiated procurement*, relying on *competitive proposals*
3. *Two-Step Sealed Bidding*, a combination of *bidding* and *negotiating*

Sealed Bidding (Sealed Bids)

Sealed bidding is a competitive procurement method used when the best value is expected to result from a selection of the lowest evaluated priced offer. It relies on a solicitation document called *invitation for bids* (IFB) (also known as *invitation to bid* [ITB], or *invitation to tender* [ITT]) and involves the following steps:

1. Preparation of invitation for bids
2. Publicizing of invitation for bids
3. Receipt of bids
4. Public opening of bids
5. Evaluation and comparison of bids
6. Selection of the lowest-priced technically acceptable solution
7. Award of contract

Sealed bids are normally solicited if

- Time permits the solicitation, submission, and evaluation of sealed bids
- The award will be made on the basis of price and other price-related factors

- It is not necessary to conduct discussions with the responding providers about their bids
- There is a reasonable expectation of receiving more than one sealed bid

Negotiated Procurement (Competitive Proposals)

Negotiated procurement is a competitive procurement method used when the best value is expected to result from selection of technically acceptable proposals, with the lowest evaluated price, in other words, when cost is not the most important factor of evaluation.

The negotiated procurement method relies on a solicitation document called *request for proposals* (RFP).

This process involves the following steps:

1. Preparation of RFP
2. Release of a public notice of solicitation
3. Receipt of *letters of intent* and *no-bid letters*
4. Preproposal meeting, mandatory or optional
5. Receipt of *proposals* and *proposal cover letters*, kept closed in a secure place until due date
6. Addenda or amendments to the RFP and, eventually, extension of the proposal receipt due date
7. Modification of the initial RFP, modification and receipt of proposals, and, eventually, extension of the proposal receipt due date
8. *Disqualification of proposals*, proposals returned unopened
9. Proposal opening at proposal receipt due date
10. Rating, scoring, and sorting proposals in a *decision matrix*
11. Selection of the best matching proposals
12. *Reject nonresponsive proposals* or *nonresponsible providers*
13. Providers contacted and requested for their best and final offer (BAFO)
14. Selection of the best matching proposal
15. *Decline unsuccessful proposal*
16. Handling of *protest letters*
17. *Award notice*

Competitive proposals are normally solicited when

- The use of sealed bids is not deemed appropriate
- It is necessary to conduct discussions with providers, because of differences in areas such as law, regulations, and business practices, especially for contracts to be made abroad

Two-Step Sealed Bidding

Two-step sealed bidding is a competitive procurement method that is a combination of the two aforementioned competitive procurement procedures and is designed to obtain

- The benefits of negotiation for helping the requesting organization complete or define specifications without any pricing consideration
- The benefits of sealed bidding for getting the best price for the technically acceptable solution agreed upon

Two-step sealed bids can be used in preference to negotiated procurement when all of the following conditions are present:

- Specifications are not definite or complete or may be, without further technical evaluation or discussion, too restrictive to ensure mutual understanding between each source and the requesting organization
- Definite criteria exist for evaluating technical proposals
- More than one technically qualified source is expected to be available
- Sufficient time will be available for use of the two-step sealed-bidding method
- A firm-fixed-price contract or a fixed-price contract with economic price adjustment will be used

Other than Full and Open Competition (OFOC)

Because Part 6 of the FAR, formerly the *Competition in Contracting Act* (CICA), defines competition as the standard for procurement. Other noncompetitive procurement methods are considered to be exceptions; thus, they should be used only under certain, well-defined conditions, and they should be carefully and thoroughly justified and documented. Contracting without providing for *full and open competition* is called *other than full and open competition* (OFOC). OFOC cannot be justified because of a failure to plan in advance or concerns about the availability of related funds or budget.

Sole source is an example of noncompetitive procurement or purchase process accomplished after soliciting and negotiating with only one source, sole source, thus limiting full and open competition. Sole source solicitation constitutes a violation of the CICA unless it is justified under one of seven specific circumstances. Read further information about *how to protest against sole source solicitation*.

These circumstances, which can be invoked as sole source justification, are described in FAR 6.302 as follows:

1. Only one responsible source and no other supplies or services will satisfy requirements
2. Unusual and compelling urgency

3. Industrial mobilization; engineering, developmental, or research capability; or expert services
4. International agreement
5. Authorized or required by statute
6. National security
7. Public interest

Although only one of these conditions may justify setting competition aside, each presents limitations that are supposed to hamper, or limit, their abuse.

If a sole source acquisition is ethical, it can shorten the acquisition process, and lead to several benefits:

- A reduced decision cycle (in other words, less time is required to award contracts)
- Lowered administrative costs (the procurement process is reduced to a minimum)
- Prices reflecting the best value
- Better promoted small business participation
- Improved delivery of products and services

Conclusion

As *Terence* said in his play, *homo sum, humani a mi nihil alienum puto* (I am a human being, so nothing human is strange to me). Humanity is, above all, our nature, and commerce and trade have always been accompanied by some degree of faulty business ethics. Yet, FAR has been successful and is inspiring more and more private enterprises to exercise due diligence in their procurement practices. By maintaining a healthy level of competition, both the buying and supplying organizations will benefit.

Web Resources on Competitive Procurement

The Project on Government Oversight

The Project on Government Oversight (POGO) is an independent nonprofit that investigates and exposes corruption, fraud, abuse, waste, and other misconduct in order to achieve a more accountable federal government. POGO provides an archive of investigations in many areas, a blog, and a federal contractor misconduct database. Reported facts, such as the $7,600 coffee maker and the $436 hammer, will dismay you.

> The Project on Government Oversight follows a rich tradition of assuring that the government continues to work for the people it represents.

> Our nation was founded on the very principle that representation and accountability are fundamental to maintaining a strong and functioning democracy. Today, these principles espoused by our founding fathers are under attack as our federal government is more vulnerable than ever to the influence of money in politics and powerful special interests. (www.pogo.org)

Complying with the "Made in the USA" Standard

Introduction

The Federal Trade Commission (FTC) is charged with preventing deception and unfairness in the marketplace. The FTC Act gives the Commission the power to bring law enforcement actions against false or misleading claims that a product is of U.S. origin. Traditionally, the Commission has required that a product advertised as Made in USA be "all or virtually all" made in the United States. After a comprehensive review of Made in USA and other U.S. origin claims in product advertising and labeling, the Commission announced in December 1997 that it would retain the "all or virtually all" standard. The Commission also issued an Enforcement Policy Statement on U.S. Origin Claims to provide guidance to marketers who want to make an unqualified Made in USA claim under the "all or virtually all" standard and those who want to make a qualified Made in USA claim.

This publication provides additional guidance about how to comply with the "all or virtually all" standard. It also offers some general information about the U.S. Customs Service's requirement that all products of foreign origin imported into the United States be marked with the name of the country of origin.

This publication is the Federal Trade Commission staff's view of the law's requirements. It is not binding on the Commission. The Enforcement Policy Statement issued by the FTC is at the end of the publication.

Basic Information about Made in USA Claims

Must U.S. content be disclosed on products sold in the U.S.?

U.S. content must be disclosed on automobiles and textile, wool, and fur products. There's no law that requires most other products sold in the U.S. to be marked or labeled Made in USA or have any other disclosure about their amount of U.S. content. However, manufacturers and marketers who choose to make claims about the amount of U.S. content in their products must comply with the FTC's Made in USA policy.

What products does the FTC's Made in USA policy apply to?

The policy applies to all products advertised or sold in the U.S., except for those specifically subject to country-of-origin labeling by other laws. Other countries may have their own country-of-origin marking requirements. As a result, exporters should determine whether the country to which they are exporting imposes such requirements.

What kinds of claims does the Enforcement Policy Statement apply to?

The Enforcement Policy Statement applies to U.S. origin claims that appear on products and labeling, advertising, and other promotional materials. It also applies to all other forms of marketing, including marketing through digital or electronic mechanisms, such as Internet or e-mail.

A Made in USA claim can be express or implied.

Examples of express claims: Made in USA. "Our products are American-made." "USA."

In identifying implied claims, the Commission focuses on the overall impression of the advertising, label, or promotional material. Depending on the context, U.S. symbols or geographic references (for example, U.S. flags, outlines of U.S. maps, or references to U.S. locations of headquarters or factories) may convey a claim of U.S. origin either by themselves, or in conjunction with other phrases or images.

Example: A company promotes its product in an ad that features a manager describing the "true American quality" of the work produced at the company's American factory. Although there is no express representation that the company's product is made in the U.S., the overall—or net—impression the ad is likely to convey to consumers is that the product is of U.S. origin.

Brand names and trademarks

Ordinarily, the Commission will not consider a manufacturer or marketer's use of an American brand name or trademark by itself as a U.S. origin claim. Similarly, the Commission is not likely to interpret the mere listing of a company's U.S. address on a package label in a non-prominent way as a claim of U.S. origin.

Example: A product is manufactured abroad by a well-known U.S. company. The fact that the company is headquartered in the U.S. also is widely known. Company pamphlets for its foreign-made product prominently feature its brand name. Assuming that the brand name does not specifically denote U.S. origin (that is, the brand name is not "Made in America, Inc."), using the brand name by itself does not constitute a claim of U.S. origin.

Representations about entire product lines

Manufacturers and marketers should not indicate, either expressly or implicitly, that a whole product line is of U.S. origin ("Our products are made in USA") when only some products in the product line are made in the U.S. according to the "all or virtually all" standard.

Does the FTC pre-approve Made in USA claims?

The Commission does not pre-approve advertising or labeling claims. A company doesn't need approval from the Commission before making a Made in USA claim. As with most other advertising claims, a manufacturer or marketer may make any claim as long as it is truthful and substantiated.

The Standard for Unqualified Made In USA Claims

What is the standard for a product to be called Made in USA without qualification?

For a product to be called Made in USA, or claimed to be of domestic origin without qualifications or limits on the claim, the product must be "all or virtually all" made in the U.S. The term "United States," as referred to in the Enforcement Policy Statement, includes the 50 states, the District of Columbia, and the U.S. territories and possessions.

What does "all or virtually all" mean?

"All or virtually all" means that all significant parts and processing that go into the product must be of U.S. origin. That is, the product should contain no—or negligible—foreign content.

What substantiation is required for a Made in USA claim?

When a manufacturer or marketer makes an unqualified claim that a product is Made in USA, it should have—and rely on—a "reasonable basis" to support the claim at the time it is made. This means a manufacturer or marketer needs competent and reliable evidence to back up the claim that its product is "all or virtually all" made in the U.S.

What factors does the Commission consider to determine whether a product is "all or virtually all" made in the U.S.?

The product's final assembly or processing must take place in the U.S. The Commission then considers other factors, including how much of the product's total manufacturing costs can be assigned to U.S. parts and processing, and how far removed any foreign content is from the finished product. In some instances, only a small portion of the total manufacturing costs are attributable to foreign processing, but that processing represents a significant amount of the product's overall processing. The same could be true for some foreign parts. In these cases, the foreign content (processing or parts) is more than negligible, and, as a result, unqualified claims are inappropriate.

Example: A company produces propane barbecue grills at a plant in Nevada. The product's major components include the gas valve, burner and aluminum housing, each of which is made in the U.S. The grill's knobs and tubing are imported from Mexico. An unqualified Made in USA claim is not likely to be deceptive because the knobs and tubing make up a negligible portion of the product's total manufacturing costs and are insignificant parts of the final product.

Example: A table lamp is assembled in the U.S. from American-made brass, an American-made Tiffany-style lampshade, and an imported base. The base accounts for a small percent of the total cost of making the lamp. An unqualified Made in USA claim is deceptive for two reasons: The base is not far enough removed in the manufacturing process from the finished product to be of little consequence and it is a significant part of the final product.

What items should manufacturers and marketers include in analyzing the percentage of domestic content in a particular product?

Manufacturers and marketers should use the cost of goods sold or inventory costs of finished goods in their analysis. Such costs generally are limited to the total cost of all manufacturing materials, direct manufacturing labor, and manufacturing overhead.

Should manufacturers and marketers rely on information from American suppliers about the amount of domestic content in the parts, components, and other elements they buy and use for their final products?

If given in good faith, manufacturers and marketers can rely on information from suppliers about the domestic content in the parts, components, and other elements they produce. Rather than assume that the input is 100 percent U.S.-made, however, manufacturers and marketers would be wise to ask the supplier for specific information about the percentage of U.S. content before they make a U.S. origin claim.

Example: A company manufactures food processors in its U.S. plant, making most of the parts, including the housing and blade, from U.S. materials. The motor, which constitutes 50 percent of the food processor's total manufacturing costs, is bought from a U.S. supplier. The food processor manufacturer knows that the motor is assembled in a U.S. factory. Even though most of the parts of the food processor are of U.S. origin, the final assembly is in the U.S., and the motor is assembled in the U.S., the food processor is not considered "all or virtually all" American-made if the motor itself is made of imported parts that constitute a significant percentage of the appliance's total manufacturing cost. Before claiming the product is Made in USA, this manufacturer should look to its motor supplier for more specific information about the motor's origin.

Example: On its purchase order, a company states: "Our company requires that suppliers certify the percentage of U.S. content in products supplied to us. If you are unable or unwilling to make such certification, we will not purchase from you." Appearing under this statement is the sentence, `"We certify that our ___ have at least ___% U.S. content," with space for the supplier to fill in the name of the product and its percentage of U.S. content. The company generally could rely on a certification like this to determine the appropriate country-of-origin designation for its product.

How far back in the manufacturing process should manufacturers and marketers look?

To determine the percentage of U.S. content, manufacturers and marketers should look back far enough in the manufacturing process to be reasonably sure that any significant foreign content has been included in their assessment of foreign costs. Foreign content incorporated early in the manufacturing process often will be less significant to consumers than content that is a direct part of the finished product or the parts or components produced by the immediate supplier.

Example: The steel used to make a single component of a complex product (for example, the steel used in the case of a computer's floppy drive) is an early input into the computer's manufacture, and is likely to constitute a very small portion of the final product's total cost. On the other hand, the steel in a product like a pipe or a wrench is a direct and significant input. Whether the steel in a pipe or wrench is imported would be a significant factor in evaluating whether the finished product is "all or virtually all" made in the U.S.

Are raw materials included in the evaluation of whether a product is "all or virtually all" made in the U.S.?

It depends on how much of the product's cost the raw materials make up and how far removed from the finished product they are.

Example: If the gold in a gold ring is imported, an unqualified Made in USA claim for the ring is deceptive. That's because of the significant value the gold is likely to represent relative to the finished product, and because the gold—an integral component—is only one step back from the finished article. By contrast, consider the plastic in the plastic case of a clock radio otherwise made in the U.S. of U.S.-made components. If the plastic case was made from imported petroleum, a Made in USA claim is likely to be appropriate because the petroleum is far enough removed from the finished product, and is an insignificant part of it as well.

Qualified Claims

What is a qualified Made in USA claim?

A qualified Made in USA claim describes the extent, amount or type of a product's domestic content or processing; it indicates that the product isn't entirely of domestic origin.

Example: "60% U.S. content." "Made in USA of U.S. and imported parts." "Couch assembled in USA from Italian Leather and Mexican Frame."

When is a qualified Made in USA claim appropriate?

A qualified Made in USA claim is appropriate for products that include U.S. content or processing but don't meet the criteria for making an unqualified Made in USA claim. Because even qualified claims may imply more domestic content than exists, manufacturers or marketers must exercise care when making these claims. That is, avoid qualified claims unless the product has a significant amount of U.S. content or U.S. processing. A qualified Made in USA claim, like an unqualified claim, must be truthful and substantiated.

Example: An exercise treadmill is assembled in the U.S. The assembly represents significant work and constitutes a "substantial transformation" (a term used by the U.S. Customs Service). All of the treadmill's major parts, including the motor, frame, and electronic display, are imported. A few of its incidental parts, such as the handle bar covers, the plastic on/off power key, and the treadmill mat, are manufactured in the U.S. Together, these parts account for approximately three percent of the total cost of all the parts. Because the value of the U.S.-made parts is negligible compared to the value of all the parts, a claim on the treadmill that it is "Made in USA of U.S. and Imported Parts" is deceptive. A claim like "Made in U.S. from Imported Parts" or "Assembled in USA" would not be deceptive.

U.S. origin claims for specific processes or parts

Claims that a particular manufacturing or other process was performed in the U.S. or that a particular part was manufactured in the U.S. must be truthful, substantiated, and clearly refer to the specific process or part, not to the general manufacture of the product, to avoid implying more U.S. content than exists.

Manufacturers and marketers should be cautious about using general terms, such as "produced," "created" or "manufactured" in the U.S. Words like these are unlikely to convey a message limited to a particular process. Additional qualification probably is necessary to describe a product that is not "all or virtually all" made in the U.S.

In addition, if a product is of foreign origin (that is, it has been substantially transformed abroad), manufacturers and marketers also should make sure they satisfy Customs' markings statute and regulations that require such products to be marked with a foreign country of origin. Further, Customs requires the foreign country of origin to be preceded by "Made in," "Product of," or words of similar meaning when any city or location that is not the country of origin appears on the product.

Example: A company designs a product in New York City and sends the blueprint to a factory in Finland for manufacturing. It labels the product "Designed in USA—Made in Finland." Such a specific processing claim would not lead a reasonable consumer to believe that the whole product was made in the U.S. The Customs Service requires the product to be marked "Made in," or "Product of" Finland since the product is of Finnish origin and the claim refers to the U.S. Examples of other specific processing claims are: "Bound in U.S.—Printed in Turkey." "Hand carved in U.S.—Wood from Philippines." "Software written in U.S.—Disk made in India." "Painted and fired in USA. Blanks made in (foreign country of origin)."

Example: A company advertises its product, which was invented in Seattle and manufactured in Bangladesh, as "Created in USA." This claim is deceptive because consumers are likely to interpret the term "Created" as Made in USA—an unqualified U.S. origin claim.

Example: A computer imported from Korea is packaged in the U.S. in an American-made corrugated paperboard box containing only domestic materials and domestically produced expanded rigid polystyrene plastic packing. Stating Made in USA on the package would deceive consumers about the origin of the product inside. But the company could legitimately make a qualified claim, such as "Computer Made in Korea—Packaging Made in USA."

Example: The Acme Camera Company assembles its cameras in the U.S. The camera lenses are manufactured in the U.S., but most of the remaining parts are imported. A magazine ad for the camera is headlined "Beware of Imported Imitations" and states "Other high-end camera makers use imported parts made with cheap foreign labor.

But at Acme Camera, we want only the highest quality parts for our cameras and we believe in employing American workers. That's why we make all of our lenses right here in the U.S." This ad is likely to convey that more than a specific product part (the lens) is of U.S. origin. The marketer should be prepared to substantiate the broader U.S. origin claim conveyed to consumers viewing the ad.

Comparative Claims

Comparative claims should be truthful and substantiated, and presented in a way that makes the basis for comparison clear (for example, whether the comparison is to another leading brand or to a previous version of the same product). They should truthfully describe the U.S. content of the product and be based on a meaningful difference in U.S. content between the compared products.

Example: An ad for cellular phones states "We use more U.S. content than any other cellular phone manufacturer." The manufacturer assembles the phones in the U.S. from American and imported components and can substantiate that the difference between the U.S. content of its phones and that of the other manufacturers' phones is significant. This comparative claim is not deceptive.

Example: A product is advertised as having "twice as much U.S. content as before." The U.S. content in the product has been increased from 2 percent in the previous version to 4 percent in the current version. This comparative claim is deceptive because the difference between the U.S. content in the current and previous version of the product are insignificant.

Assembled in USA Claims

A product that includes foreign components may be called "Assembled in USA" without qualification when its principal assembly takes place in the U.S. and the assembly is substantial. For the "assembly" claim to be valid, the product's last "substantial transformation" also should have occurred in the U.S. That's why a "screwdriver" assembly in the U.S. of foreign components into a final product at the end of the manufacturing process doesn't usually qualify for the "Assembled in USA" claim.

Example: A lawn mower, composed of all domestic parts except for the cable sheathing, flywheel, wheel rims and air filter (15 to 20 percent foreign content) is assembled in the U.S. An "Assembled in USA" claim is appropriate.

Example: All the major components of a computer, including the motherboard and hard drive, are imported. The computer's components then are put together in a simple "screwdriver" operation in the U.S., are not substantially transformed under the Customs Standard, and must be marked with a foreign country of origin. An "Assembled in U.S." claim without further qualification is deceptive.

The FTC and the Customs Service

What is the U.S. Customs Service's jurisdiction over country-of-origin claims?

The Tariff Act gives Customs and the Secretary of the Treasury the power to administer the requirement that imported goods be marked with a foreign country of origin (for example, "Made in Japan").

When an imported product incorporates materials and/or processing from more than one country, Customs considers the country of origin to be the last country in which a "substantial transformation" took place. Customs defines "substantial transformation" as a manufacturing process that results in a new and different product with a new name, character, and use that is different from that which existed before the change. Customs makes country-of-origin determinations using the "substantial transformation" test on a case-by-case basis. In some instances, Customs uses a "tariff shift" analysis, comparable to "substantial transformation," to determine a product's country of origin.

What is the interaction between the FTC and Customs regarding country-of-origin claims?

Even if Customs determines that an imported product does not need a foreign country-of-origin mark, it is not necessarily permissible to promote that product as Made in USA. The FTC considers additional factors to decide whether a product can be advertised or labeled as Made in USA.

Manufacturers and marketers should check with Customs to see if they need to mark their products with the foreign country of origin. If they don't, they should look at the FTC's standard to check if they can properly make a Made in USA claim.

The FTC has jurisdiction over foreign origin claims on products and in packaging that are beyond the disclosures required by Customs (for example, claims that supplement a required foreign origin marking to indicate where additional processing or finishing of a product occurred).

The FTC also has jurisdiction over foreign origin claims in advertising and other promotional materials. Unqualified U.S. origin claims in ads or other promotional materials for products that Customs requires a foreign country-of-origin mark may mislead or confuse consumers about the product's origin. To avoid misleading consumers, marketers should clearly disclose the foreign manufacture of a product.

Example: A television set assembled in Korea using an American-made picture tube is shipped to the U.S. The Customs Service requires the television set to be marked "Made in Korea" because that's where the television set was last "substantially transformed." The company's World Wide Web page states "Although our televisions are made abroad, they always contain U.S.-made picture tubes." This statement is not deceptive. However, making the statement "All our picture tubes are made in the USA"—without disclosing the foreign origin of the television's manufacture—might imply a broader claim (for example, that the television set is largely made in the U.S.) than could be substantiated. That is, if the statement and the entire ad imply that any foreign content or processing is negligible, the advertiser must substantiate that claim or net impression. The advertiser in this scenario would not be able to substantiate the implied Made in USA claim because the product was "substantially transformed" in Korea.

Other Statutes

What are the requirements of other federal statutes relating to country-of-origin determinations?

Textile Fiber Products Identification Act and Wool Products Labeling Act—Require a Made in USA label on most clothing and other textile or wool household products if the final product is manufactured in the U.S. of fabric that is manufactured in the U.S., regardless of where materials earlier in the manufacturing process (for example, the yarn and fiber) came from. Textile products that are imported must be labeled as required by the Customs Service. A textile or wool product partially manufactured in the U.S. and partially manufactured in another country must be labeled to show both foreign and domestic processing.

On a garment with a neck, the country of origin must be disclosed on the front of a label attached to the inside center of the neck—either midway between the shoulder seams or very near another label attached to the inside center of the neck. On a garment without a neck, and on other kinds of textile products, the country of origin must appear on a conspicuous and readily accessible label on the inside or outside of the product.

Catalogs and other mail order promotional materials for textile and wool products, including those disseminated on the Internet, must disclose whether a product is made in the U.S., imported or both.

The Fur Products Labeling Act requires the country of origin of imported furs to be disclosed on all labels and in all advertising. For copies of the Textile, Wool or Fur Rules and Regulations, or the new business education guide on labeling requirements, call the FTC's Consumer Response Center (202-382-4357). Or visit the FTC online at www.ftc.gov. Click on Consumer Protection.

American Automobile Labeling Act—Requires that each automobile manufactured on or after October 1, 1994, for sale in the U.S. bear a label disclosing where the car was assembled, the percentage of equipment that originated in the U.S. and Canada, and the country of origin of the engine and transmission. Any representation that a car marketer makes that is required by the AALA is exempt from the Commission's policy. When a company makes claims in advertising or promotional materials that go beyond the AALA requirements, it will be held to the Commission's standard. For more information, call the Consumer Programs Division of the National Highway Traffic Safety Administration (202-366-0846).

Buy American Act—Requires that a product be manufactured in the U.S. of more than 50 percent U.S. parts to be considered Made in USA for government procurement purposes. For more information, review the Buy American Act at 41 U.S.C. §§ 10a-10c, the Federal Acquisition Regulations at 48 C.F.R. Part 25, and the Trade Agreements Act at 19 U.S.C. §§ 2501-2582.

What to Do about Violations

What if I suspect noncompliance with the FTC's Made in USA standard or other country-of-origin mislabeling?

Information about possible illegal activity helps law enforcement officials target companies whose practices warrant scrutiny. If you suspect noncompliance, contact the Division of Enforcement, Bureau of Consumer Protection, Federal Trade Commission, Washington, DC 20580; (202) 326-2996 or send an e-mail to MUSA@ftc.gov. If you know about import or export fraud, call Customs' toll-free Commercial Fraud Hotline, 1-800-ITS-FAKE. Examples of fraudulent practices involving imports include removing a required foreign origin label before the product is delivered to the ultimate purchaser (with or without the improper substitution of a Made in USA label) and failing to label a product with a required country of origin.

You also can contact your state attorney general and your local Better Business Bureau to report a company, or you can refer your complaint to the National Advertising Division (NAD) of the Council of Better Business Bureaus by calling (212) 754-1320. NAD handles complaints about the truth and accuracy of national advertising. You can reach the Council of Better Business Bureaus on the Web at adweb.com/adassoc17.html.

Finally, the Lanham Act gives any person (such as a competitor) who is damaged by a false designation of origin the right to sue the party making the false claim. Consult a lawyer to see if this private right of action is an appropriate course of action for you.

For More Information

The FTC works for the consumer to prevent fraudulent, deceptive, and unfair practices in the marketplace and to provide information to businesses to help them comply with the law. To file a complaint or to get free information on consumer issues, visit www.ftc.gov or call toll-free, 1-877-FTC-HELP (1-877-382-4357); TTY: 1-866-653-4261. The FTC enters Internet, telemarketing, identity theft, and other fraud-related complaints into Consumer Sentinel, a secure online database available to hundreds of civil and criminal law enforcement agencies in the U.S. and abroad.

Your Opportunity to Comment

The National Small Business Ombudsman and 10 Regional Fairness Boards collect comments from small businesses about federal compliance and enforcement activities. Each year, the Ombudsman evaluates the conduct of these activities and rates each agency's responsiveness to small businesses. Small businesses can comment to the Ombudsman without fear of reprisal. To comment, call toll-free 1-888-REGFAIR (1-888-734-3247) or go to www.sba.gov/ombudsman.

Recall Handbook

A Guide for Manufacturers, Importers, Distributors and Retailers on Reporting Under Sections 15 and 37 of the Consumer Product Safety Act and Section 102 of the Child Safety Protection Act and Preparing for, Initiating and Implementing Product Safety Recalls

Including:

CPSC Fast Track Product Recall Program
U.S. Consumer Product Safety Commission
Office of Compliance
Recalls and Compliance Division
4330 East West Highway, Room 613
Bethesda, Maryland 20814
Telephone: (301) 504-7913
Fax: (301) 504-0359
www.cpsc.gov

CPSC FAX-ON-DEMAND DOCUMENT #8002

May 1999

Foreword

The CPSC Office of Compliance staff prepared this *Recall Handbook* to assist your company to understand your obligations and responsibilities under the Consumer Product Safety Act. It applies to you if you manufacture, import, distribute, or retail consumer products.

No company likes to recall one of its products, but when a safety problem makes a product recall necessary to prevent injuries and save lives, it is to everyone's benefit to move quickly and effectively.

Our staff is constantly striving to improve both the timeliness of recalls and the effectiveness of the recall programs negotiated with companies. Our Fast Track Product Recall Program is helping both of these efforts and is highlighted in this revised edition of the *Recall Handbook*.

The Fast Track program is designed for companies willing and able to move quickly with a voluntary recall of their product. The program, described in detail in Section IV, eliminates some of the procedural steps in the traditional recall process, including the staff preliminary determination that the product contains a defect that presents a substantial product hazard.

Many companies have used the program since CPSC introduced it in August 1995. Now about half of our recalls are done under this streamlined system. In 1998 the CPSC Fast Track Product Recall program was one of three federal winners of the prestigious Innovations in American Government Award. This award is funded by the Ford Foundation and administered by the John F. Kennedy School of Government at Harvard University in partnership with the Council for Excellence in Government.

We welcome your comments on the Fast Track program or any other information in this handbook.

Marc J. Schoem

Director

Recalls and Compliance Division

Background

The U.S. Consumer Product Safety Commission (CPSC) is an independent regulatory agency responsible for protecting the public from unreasonable risks of injury and death associated with consumer products. Established by Congress in the Consumer Product Safety Act (CPSA), 15 U.S.C. §§ 2051-2084, the CPSC has jurisdiction over approximately 15,000 different types of products used in and around the home, in schools, and in recreation ("consumer products").[*][†]

This handbook is for companies that manufacture, import, distribute, retail, or otherwise sell consumer products. It has three purposes: 1) to familiarize companies with their reporting requirements under sections 15(b) and 37 of the CPSA, 15 U.S.C. § 2064(b) and § 2084, and Section 102 of the Child Safety Protection Act, Pub. L. 103-267, 108 Stat. 722, 6/16/94; 2) to help companies learn how to recognize potentially hazardous consumer products at an early stage; and 3) to assist firms that discover they have manufactured or distributed such products develop and implement "corrective action plans" that address the hazards. The term "corrective action plan" (CAP) generally includes any type of remedial action taken by a firm. A CAP could, for example, provide for the return of a product to the manufacturer or retailer for a cash refund or a replacement product; for the repair of a product; and/or for public notice of the hazard. A CAP may include multiple measures that are necessary to protect consumers. The Commission staff refers to corrective actions as "recalls" because the public and media more readily recognize and respond to that description.[‡]

This handbook is not an all-inclusive reference source of information describing how to recall products. The goal of a corrective action plan should be to retrieve as many hazardous products from the distribution

---

[*] This handbook does not replace the Commission's statutes or interpretative regulations set out in 16 C.F.R. Parts 1115, 1116 and 1117. If there is any discrepancy, the statutes and regulations supersede this handbook. For more information about reporting requirements, see the Commission's Statement of Enforcement Policy, 51 Fed. Reg. 23,410 (1986). This material is available on the CPSC Web site at: http://www.cpsc.gov.

[†] The Commission does not have general jurisdiction over foods, drugs, cosmetics, medical devices, firearms and ammunition, boats, motor vehicles, aircraft, or tobacco. Specific questions about the Commission's jurisdiction over particular products should be directed to the Office of the General Counsel.

[‡] This handbook uses the term "recall" to describe any repair, replacement or refund program.

chain and from consumers as is possible in the most practical, cost-effective manner. Reaching this goal often requires creative planning. Companies developing specific corrective action plans to address unsafe or potentially unsafe products typically work closely with the Commission staff to take advantage of the staff's expertise in designing and carrying out such plans. This results in greater protection for consumers against injury or death.

I. Reporting Requirements.

A. Section 15 Reports

Section 15(b) of the Consumer Product Safety Act establishes reporting requirements for manufacturers, importers, distributors and retailers of consumer products. Each must notify the Commission immediately if it obtains information which reasonably supports the conclusion that a product distributed in commerce (1) fails to meet a consumer product safety standard or banning regulation, (2) contains a defect which could create a substantial product hazard to consumers, (3) creates an unreasonable risk of serious injury or death, or (4) fails to comply with a voluntary standard upon which the Commission has relied under the CPSA.* Companies that distribute products that violate regulations issued under the other laws that the Commission administers—the Flammable Fabrics Act, 15 U.S.C. § 1193-1204; the Federal Hazardous Substances Act, 15 U.S.C. § 1261-1278; the Poison Prevention Packaging Act, 15 U.S.C. § 1471-1476; and the Refrigerator Safety Act; 15 U.S.C. §1211-1214—must also report, if the violations may also constitute product defects that could create a substantial risk of injury to the public or may create an unreasonable risk of serious injury or death. The Commission has issued an interpretive regulation, 16 C.F.R. Part 1115, that further explains a reporting company's obligations.

In enacting section 15(b), Congress intended to encourage the widespread reporting of potential product hazards. In addition to assisting the Commission to discover substantial product hazards, reporting would identify risks of injury that the Commission could address through voluntary or mandatory standards, or information and education.

Although CPSC uses sources other than company reports to identify potentially hazardous products, reporting by companies under section 15 can provide the most timely and effective source of information

* As of October 1998, there were two such standards—the voluntary standards for chain saws and for unvented gas space heaters.

about such products. This is because firms often learn of potential product safety problems at an early stage. For this reason, companies involved in the manufacture, importation, distribution, or sale of consumer products should develop a system for maintaining and reviewing information about their products that might suggest a product defect or unreasonable risk of serious injury or death. Such information includes consumer complaints, warranty returns, insurance claims or payments, product liability lawsuits, reports of production problems, product testing or other critical analyses of products, and the like.

Reporting a product to the Commission under section 15 does not automatically mean that the Commission will conclude that the product creates a substantial product hazard or that corrective action is necessary. The CPSC staff works with the reporting firm to determine if corrective action is appropriate. Many of the reports received require no corrective action because the staff concludes that the reported product defect does not create a substantial product hazard.

1. What and Where to Report

A company should file its report with the Division of Recalls and Compliance. The report may be filed by mail, telephone (301-504-7913), or electronically through the CPSC Web site (www.cpsc.gov) or fax (301-504-0359). A company should assign the responsibility of reporting to someone with knowledge of the product and of the reporting requirements of section 15. He or she should have the authority to report to CPSC or to quickly raise the reporting issue to someone who does.

Reporting firms should be prepared to provide the information described below. However, no company should delay a report because some of this information is not yet available. The following information should be transmitted:

- Description of the product.
- Name and address of the company, and whether it is a manufacturer, distributor, importer or retailer.
- Nature and extent of the possible product defect or unreasonable risk of serious injury or death.
- Nature and extent of injury or possible injury associated with the product.
- Name, address and telephone number of the person informing the Commission.
- If available, the other information specified in Section 1115.13(d) of the Commission's regulations interpreting the reporting requirements.

- A timetable for providing information not immediately available.

Retailers and distributors may satisfy their reporting obligations in the manner described above. Alternatively, a retailer or distributor may send a letter to the manufacturer or importer of a product describing the risk of injury or death or the defect associated with the product or its failure to comply with an applicable regulation and forward a copy of that letter to the Division of Recalls and Compliance. A distributor or retailer receiving product hazard information from a manufacturer or importer or other source must report to CPSC unless the firm knows the Commission has been adequately informed of the defect, failure to comply, or risk.

2. When to Report

Section 15 requires firms to report "immediately." This means that a firm should notify the Commission within 24 hours of obtaining information described in section A ("Section 15 Reports") above. 16 C.F.R. § 1115.12 provides guidelines for determining whether a product defect exists, whether a product creates an unreasonable risk of serious injury or death, and whether a report is necessary or appropriate. Section II of this handbook does the same.

A company *must* report to the Commission within 24 hours of obtaining reportable information. The Commission encourages companies to report *potential* substantial product hazards even while their own investigations are continuing. However, if a company is uncertain whether information is reportable, the firm may spend a reasonable time investigating the matter. That investigation should not exceed ten working days unless the firm can demonstrate that a longer time is reasonable in the circumstances. Absent such circumstances, the Commission will presume that, at the end of ten working days, the firm has received and considered all information which would have been available to it had a reasonable, expeditious, and diligent investigation been undertaken.

The Commission considers a company to have obtained knowledge of product safety related information when that information is received by an employee or official of the firm who may reasonably be expected to be capable of appreciating the significance of that information. Once that occurs, under ordinary circumstances, five working days is the maximum reasonable time for that information to reach the chief executive officer or the official assigned responsibility for complying with the reporting requirements.

The Commission evaluates whether or when a firm should have reported. This evaluation will be based, in part, on what the company actually knew about the hazard posed by the product or *on what a reasonable person, acting under the circumstances, should have known about the hazard while exercising due care.* Thus, a firm is deemed to know what it would have known had it exercised due care in analyzing reports of injury or consumer complaints, or in evaluating warranty returns, reports of experts, in-house engineering analyses, or other information.

3. Confidentiality of Reports

The Commission often receives requests for information reported under section 15(b). Section 6(b)(5) of the CPSA, 15 U.S.C. § 2055(b)(5), prohibits the release of such information unless a remedial action plan has been accepted in writing, a complaint has been issued or the reporting firm consents to the release. In addition, a firm claiming that information it has submitted is trade secret or confidential commercial or financial information must mark the information as "confidential" in accordance with section 6(a)(3) of the CPSA, 15 U.S.C. § 2055(a)(3). That should be done when the information is submitted to the Commission. The firm will receive an additional opportunity to claim confidentiality when it receives subsequent notice from the Commission's Freedom of Information Office that the information may be disclosed to the public in response to a request. If section 6(b)(5) does not apply, the CPSC staff will not treat information as exempt from disclosure to the public under section 6(a) of the CPSA, 15 U.S.C. § 2055(a), and the Freedom of Information Act, absent a specific claim for confidential treatment.

B. Section 37 Reports

Section 37 of the CPSA requires manufacturers of consumer products to report information about settled or adjudicated lawsuits.* Manufacturers must report if:

- A particular model of the product is the subject of at least three civil actions filed in federal or state court;

* The Commission has issued a rule interpreting the requirements of section 37 at 16 C.F.R. Part 1116. The Commission recommends that manufacturers considering whether they have section 37 reporting obligations refer to that rule, particularly in determining whether products involved in different lawsuits are the same model.

- Each suit alleges the involvement of that particular model in death or grievous bodily injury—mutilation or disfigurement, dismemberment or amputation, the loss of important bodily functions or debilitating internal disorder, injuries likely to require extended hospitalization, severe burns, severe electric shock, or other injuries of similar severity; and
- During one of the following two-year periods specified in the law, each of the three actions results in either a final settlement involving the manufacturer or in a court judgment in favor of the plaintiff:
  - January 1, 1991–December 31, 1992
  - January 1, 1993–December 31, 1994
  - January 1, 1995–December 31, 1996
  - January 1, 1997–December 31, 1998
  - January 1, 1999–December 31, 2000
  - subsequent periods follow this pattern); and
- The manufacturer is involved in the defense of or has notice of each action prior to the entry of the final order, and is involved in discharging any obligation owed to the plaintiff as a result of the settlement or judgment.

1. What to Report

A report under section 37 must contain:

- The name and address of the manufacturer of the product.
- The model and model number or designation of the product.
- A statement as to whether the civil action alleged death or grievous bodily injury, and in the case of the latter, the nature of the injury. For reporting purposes, the plaintiff's allegations as to the nature of the injury are sufficient to require a report, even if the manufacturer disagrees with the allegations.
- A statement as to whether the case resulted in a final settlement or a judgment. However, a manufacturer need not provide the amount of a settlement.
- In the case of a judgment in favor of the plaintiff, the name and case number of the case, and the court in which it was filed.

A manufacturer may also provide additional information, if it chooses. Such information might include a statement as to whether the manufacturer intends to appeal an adverse judgment, a specific denial that the information it submits reasonably supports the conclusion that its product caused death or grievous bodily injury, and an explanation why the manufacturer has not previously reported the risk associated with the product under section 15.

2. When and Where to Report

A manufacturer must report within 30 days after a judgment or final settlement in the last of three lawsuits. The same is true of any additional lawsuits involving the same model that are settled or adjudicated in favor of the plaintiff during the same two-year period.

Companies must file Section 37 reports in writing to the Director, Recalls and Compliance Division, Office of Compliance, U.S. Consumer Product Safety Commission, Washington, D.C. 20207.

3. Confidentiality of Reports

Unlike section 15(b) reports, the Commission may not disclose to the general public information reported under section 37 in any circumstances. By law, reporting under section 37 is not an admission of the existence of an unreasonable risk of injury, a defect, a substantial product hazard, an imminent hazard, or any other liability under any statute or common law.

C. Section 102

Section 102 of the Child Safety Protection Act requires that companies report certain choking incidents to the Commission.* Each manufacturer, distributor, retailer, and importer of a marble, a ball with a diameter of 1.75″ or less ("small ball"), or latex balloon, or a toy or game that contains such a marble, ball, balloon, or other small part must report information that reasonably supports the conclusion:

1) that a child (regardless of age) choked on such a marble, small ball, balloon, or small part; and
2) that, as a result of the incident, the child died, suffered serious injury, ceased breathing for any length of time, or was treated by a medical professional.

1. What to Report

The report should include the name and address of the child who choked and the person who notified the firm of the incident, a detailed

* Firms should refer to 16 C.F.R. Part 1117 for more detailed information about this reporting requirement.

identification of the product, a description of the incident and any resulting injuries or medical treatment, information about any changes made to the product involved or its labeling or warnings to address the risk of choking, and the details of any public notice or other corrective action planned. Firms should refer to 16 C.F.R. Part 1117 for more detailed information about this reporting requirement.

2. When and Where to Report

Section 102 reports must be filed within 24 hours of obtaining the information.

A company may file a Section 102 report with the Division of Recalls and Compliance by mail, telephone (301-504-7913), or fax (301-504-0359). Telephone reports must be followed with a written confirmation.

3. Confidentiality of Reports

Section 102 reports receive the same confidentiality treatment as information submitted under section 15 of the CPSA.

II. Identifying a Defect

The Commission's reporting requirements provide information that assists the Commission in evaluating whether some form of remedial action is appropriate. However, in the absence of a regulation that addresses a specific risk of injury, the product in question must contain a defect that creates a substantial risk of injury to the public to warrant such remedial action. The handbook next discusses the considerations that go into determining whether a product defect exists and, if so, whether the risk presented by that defect is substantial.

A defect could be the result of a manufacturing or production error; or it could result from the design of, or the materials used in, the product. A defect could also occur in a product's contents, construction, finish, packaging, warnings, and/or instructions. (See 16 C.F.R. § 1115.4)

Not all products that present a risk of injury are defective. A kitchen knife is one such example. The blade has to be sharp to allow the consumer to cut or slice food. The knife's cutting ability is not a product defect, even though some consumers may cut themselves while using the knife.

In determining whether a risk of injury associated with a product could make the product defective, the Commission considers the following:

1. *What is the utility of the product? What is it supposed to do?*
2. *What is the nature of the injury that the product might cause?*

3. *What is the need for the product?*
4. *What is the population exposed to the product and the risk of injury?*
5. *What is the Commission's experience with the product?*
6. *Finally, what other information sheds light on the product and patterns of consumer use?*

If the information available to a company does not reasonably support the conclusion that a defect exists, the firm need not report to the Commission under the defect reporting provision of Section 15(b)(2). However, since a product may be defective even when it is designed, manufactured, and marketed exactly as intended, a company in doubt as to whether a defect exists should still report. Additionally, a firm must report if it has information indicating the product creates an unreasonable risk of serious injury or death. See 15 U.S.C. 2064(b)(3) and 16 C.F.R. § 1115.6.

If the information obtained by a company supports a conclusion that a product has a defect, the company must then consider whether the defect may be serious enough that it could create a substantial product hazard. Generally, a product could create a substantial hazard when consumers are exposed to a significant number of units or if the possible injury is serious or is likely to occur. However, because a company ordinarily does not know the extent of public exposure or the likelihood or severity of potential injury when a product defect first comes to its attention, the company should report to the Commission even if it in doubt as to whether a substantial product hazard exists.

Section 15 lists criteria for determining when a product creates a substantial product hazard. Any one of the following factors could indicate the existence of a substantial product hazard:

- **Pattern of defect.** The defect may stem from the design, composition, content, construction, finish, or packaging of a product, or from warnings and/or instructions accompanying the product. The conditions under which the defect manifests itself must also be considered in determining whether the pattern creates a substantial product hazard.
- **Number of defective products distributed in commerce.** A single defective product could be the basis for a substantial product hazard determination if an injury is likely or could be serious. By contrast, defective products posing no risk of serious injury and having little chance of causing even minor injury ordinarily would not be considered to present a substantial product hazard.
- **Severity of risk.** A risk is considered severe if the injury that might occur is serious, and/or if the injury is likely to occur.

- **Likelihood of injury.** The likelihood is determined by considering the number of injuries that have occurred, *or that could occur*, the intended or reasonably foreseeable use or misuse of the product, and the population group (such as children, the elderly, or the disabled) exposed to the product.
- A substantial product hazard also exists when a product does not comply with an applicable consumer product safety rule, and the failure to comply creates a substantial risk of injury to consumers.

III. CPSC Evaluation of Section 15 Reports

When a company reports to the Commission, the staff of the Division of Recalls and Compliance undertakes the same product hazard analysis as that requested of firms. First, the staff considers whether the product contains a defect. If the staff believes there is a defect, it then assesses the substantiality of the risk presented to the public, using the criteria listed in section 15 (that is, pattern of defect, number of defective products distributed in commerce, severity of the risk, likelihood of injury and other appropriate data). In determining preliminarily whether the product in question creates a substantial product hazard*, the staff applies hazard priority standards to classify the severity of the problem.

The hazard priority system allows the Commission staff to rank defective products uniformly. For example, a Class A hazard rating is reserved for product defects that present a strong likelihood of death or grievous injury or illness to the consumer. Should the staff make a preliminary determination that a product creates a substantial product hazard, the hazard priority system also provides a guide for selecting the level and intensity of corrective action.

Class A Hazard

Exists when a risk of death or grievous injury or illness is likely or very likely, or serious injury or illness is very likely.

Class A hazards warrant the highest level of attention. They call for a company to take immediate, comprehensive, and imaginative corrective action measures to identify and notify consumers, retailers and dis-

* The decision is preliminary because only the Commissioners, after a hearing, can make a formal determination that a product is defective and creates a substantial product hazard.

tributors having the defective product and to remedy the defect through repair or replacement of the product, refunds, or other measures.

Class B Hazard

Exists when a risk of death or grievous injury or illness is not likely to occur, but is possible, or when serious injury or illness is likely, or moderate injury or illness is very likely.

Class C Hazard

Exists when a risk of serious injury or illness is not likely, but is possible, or when moderate injury or illness is not necessarily likely, but is possible.

Regardless of whether a product defect is classified as a Class A, B or C priority hazard, the common element is that each of these defects creates a substantial product hazard that requires corrective action to reduce that risk of injury.

The priority given to a specific product defect provides a guideline for determining how best to communicate with owners and users of the defective product and to get them to respond appropriately. While some companies have exemplary track records in communicating with consumers independently, it is still to a company's advantage to work with the Commission staff, using both the company's and the Commission's skills and resources to conduct an effective product recall.

IV. Fast Track Product Recall Program

A firm that files a Section 15(b) report may wish to use of an alternative procedure that the Commission has established to expedite recalls.* The program is called the "Fast Track Product Recall Program." If a company reports a potential product defect and, within 20 working days of the filing of the report, implements with CPSC a consumer-level voluntary recall that is satisfactory to the staff, the staff will not make a preliminary determination that the product contains a defect which creates a substantial product hazard.

This program allows the staff and company to work together on a corrective action plan almost immediately, rather than spending the time and other resources necessary to investigate the reported defect further to determine whether it rises to the level of a substantial product hazard.

* This program is described in more detail in the *Federal Register* of July 24, 1997, 62 Fed. Reg. 39,827-39,828.

To participate in this program, companies must:

- Provide all of the information required for a full report (16 C.F.R. § 1115.13(d)).
- Request to participate in the program.
- Submit a proposed corrective action plan with sufficient time for the Commission staff to analyze any proposed repair, replacement or refund offer and to evaluate all notice material before the implementation (announcement) of the CAP which is to occur within 20 working days of the report.

V. Putting Together a Corrective Action Plan

A. Preparing for a Product Recall

It is unlikely that any two recall programs will ever be identical. Therefore, companies should be prepared to address issues that invariably arise. For instance:

- What is the defect that causes the product hazard?
- What caused the product defect to occur in the first place?
- Where are the unsafe products? How many are there?
- Did the product fail to comply with government safety regulations? How?
- Was the government or the appropriate regulatory body informed about the defect or lack of compliance?
- Has the company discontinued production and shipments of these products to distributors?
- Has the company notified retailers to stop selling the product and asked them to help identify consumers who own the product?
- Has the company started reviewing existing databases to identify potential product owners, e.g., product registration and customer service records?
- Has a press release been prepared announcing the recall? What other forms of public notice are needed?
- Has a toll-free telephone service been set up that will be able to handle the number of calls expected after the recall is announced?
- What is the company's estimate of the cost of the product recall campaign?
- Is the company prepared to deploy manpower and/or fund an effort to provide replacement parts for defective products or to exchange them for new products that do not have the problem?
- Has a plan been developed to ship replacement parts or new units to distributors participating in the product recall, or otherwise repair units in their inventory?

- Is the company prepared to monitor the product recall and provide timely reports to the Commission on the progress of the recall?
- How is the company upgrading its quality control or risk analysis procedures to prevent a similar product recall in the future?

This list addresses administrative and operational functions of a company involved in a product recall. Even if a product recall is merely potential, a company should be prepared to respond to the questions listed above.

B. Elements of a Recall

A company that undertakes a recall should develop a comprehensive plan that reaches throughout the entire distribution chain to consumers who have the product. The company must design each communication to motivate people to respond to the recall and take the action requested by the company.

Once the staff and a company agree on a remedy to correct a product defect, the staff works with the company to put together an effective plan for public notification and implementation of the recall. The information that should be included in a corrective action plan (CAP) is set forth at 16 C.F.R. § 1115.20(a). A plan must include the company's agreement that the Commission may publicize the terms of the plan to the extent necessary to inform the public of the nature of the alleged substantial product hazard and the actions being undertaken to correct that hazard.

The objectives of a recall are:

1. To locate all defective products as quickly as possible;
2. To remove defective products from the distribution chain and from the possession of consumers; and
3. To communicate accurate and understandable information in a timely manner to the public about the product defect, the hazard, and the corrective action. Companies should design all informational material to motivate retailers and media to get the word out and consumers to act on the recall.

In determining what forms of notice to use, the paramount consideration should be the level of hazard that the recalled product presents. Class A hazards warrant the highest level of company and Commission attention. Other considerations include where and how the product was marketed, its user population, the estimated useful life of the product, and how the product is most likely to be maintained and repaired.

A company conducting a recall must take particular care to coordinate the notice portion of the recall so that all participating parties, including retailers, have sufficient advance notice so that they can carry out the actions agreed upon. Notice also needs to be balanced—the purpose of some elements, such as news releases, press conferences, and video news releases—is to get the media to publicize information about the recall widely. Other elements, such as advertisements and posters, assure that the information is available to the public throughout the course of the recall and attempts to reach consumers who did not hear the original announcement.

VI. Communicating Recall Information

The Commission encourages companies to be creative in developing ways to reach owners of recalled products and motivate them to respond. The following are examples of types of notice that may be appropriate. This list is meant as a guide only, and is by no means all-inclusive. As new or innovative methods of notice and means of communication become available, such as use of the Internet, the staff encourages their use.

- A joint news release from CPSC and the company;
- Targeted distribution of the news release;
- A dedicated toll-free number and/or fax number for consumers to call to respond to the recall notice;
- Information on company World Wide Web sites;
- A video news release to complement the written news release;
- A national news conference and/or television or radio announcements;
- Direct notice to consumers known to have the product—identified through registration cards, sales records, catalog orders, or other means;
- Notices to distributors, dealers, sales representatives, retailers, service personnel, installers, and other persons who may have handled or been involved with the product;
- Purchase of mailing lists of populations likely to use the product;
- Paid notices via television and/or radio;
- Paid notices in national newspapers and/or magazines to reach targeted users of the product;
- Paid notices through local or regional media;
- Incentives such as money, gifts, premiums, or coupons to encourage consumers to return the product;

- Point-of-purchase posters;
- Notices in product catalogs, newsletters, and other marketing materials;
- Posters for display at locations where users are likely to visit, such as stores, medical clinics, pediatricians' offices, day care centers, repair shops, equipment rental locations, etc.;
- Notices to repair/parts shops;
- Service bulletins;
- Notices included with product replacement parts/accessories.
- Notices to day care centers.
- Notices to thrift stores.

The Compliance staff must review and agree upon each communication that a company intends to use in a product recall before publication or dissemination. It is, therefore, imperative that companies give the staff advance drafts of all notices or other communications to media, customers, and consumers.

Following are some specific suggestions for communicating recall information:

A. News Releases

Unless a company can identify all purchasers of a product being recalled and notify them directly, the Commission typically issues a news release jointly with the firm. The Compliance staff develops the wording of the release with the recalling company and in conjunction with the Commission's Office of Information and Public Affairs. The agreed-upon language for the news release provides the foundation for preparing other notice documents. The Commission discourages unilateral releases issued by companies because they create confusion among the media and public, particularly if CPSC is also issuing a release on the same subject.

The Office of Information and Public Affairs sends the news releases to national wire services, major metropolitan daily newspapers, television and radio networks, and periodicals on the agency's news contact mailing list. News releases from the Commission receive wide media attention and generate a good response rate from consumers.

Each recall news release should use the word "recall" in the heading and should begin, "In cooperation with the U.S. Consumer Product Safety Commission (CPSC)...."

Recall news releases must include the following:

- The name and location of the recalling firm
- The name of the product

- The number of products involved
- A description of the hazard
- The number of deaths, injuries, and incidents involving the product
- Detailed description of the product, including model numbers, colors, sizes, and labeling
- A line drawing or photograph of the product
- Major retailers and where and when the product was sold and retail cost
- Complete instructions for consumers on how to participate in the recall

CPSC posts recall news releases on its Internet Web site and requests companies to provide color photographs of recalled products for the web site.

B. Video News Releases

A video news release (VNR) is a taped version of the written news release that describes the recall in audio-visual terms. Distributed via satellite to television stations nationwide, it is an effective method to enhance a recall announcement. A VNR increases the chances that television news media will air information about a recall because it effectively provides news of the recall to television news producers in the form that they need.

Commission staff works with firms to produce VNRs announcing recalls. Like news releases, VNRs need to communicate basic information clearly and concisely. VNRs should incorporate the same information as the news release, as well as video images of the product. They often also include brief statements of company officials and/or the Chairman of the Commission. When writing a VNR script, remember that, if this information is to reach consumers, television networks or local stations must pick it up—which means that the script must be written for television producers.

A brief guide describing how to produce a VNR is available from the Office of Compliance upon request.

C. Posters

Posters are an effective means of providing continuing notice of recalls to consumers at points of purchase or other locations that they visit. Guidelines for posters and counter cards:

- Keep them BRIEF and eye-catching; in general, a poster requires far fewer words than a news release.
- Describe the hazard and tell consumers what to do.

- Use color to make the poster stand out.
- Use a print font, size, and color that provides a strong contrast to the background color of the poster.
- Include the terms "safety" and "recall" in the heading.
- Use a good quality line drawing or photograph of the product with call outs identifying product information, such as model numbers and date codes.
- The firm's toll-free telephone number should be in large size type at the bottom of the poster.
- The poster should include "Post until [date at least 120 days from recall announcement]."
- Consider tear-off sheets with each poster with information on the recall for consumers to take home.

A company that chooses to produce posters announcing a recall should contact the firms or individuals that the company wants to display the posters before the recall is announced. The company should explain the reason for the recall and the contribution to public safety that the posters provide. The company should also:

- Advise retailers or other firms to place the posters in several conspicuous locations in their stores or offices where customers will see them, e.g., the area where the product was originally displayed for sale, store entrances, waiting rooms in pediatric clinics, service counters at repair shops.
- Provide sufficient numbers of posters for retailers or others to display them in more than one place in each store or location, and provide a contact for ordering additional posters.

CPSC recommends that posters be 11 × 17 inches, but in no case smaller than 8.5 × 11 inches. These two sizes are easiest to mail in bulk quantities. Larger sizes may be appropriate for repair and service shops. Also, many retailers, particularly large chains, have specific requirements for posters, including size and some product identification information. To avoid delays and having to reprint, a company producing a recall poster should take care to contact retailers in advance to see if they have any such requirements.

D. Other Forms of Notice

Like news releases and posters, letters, advertisements, bulletins, newsletters, and other communications about a recall need to provide sufficient information and motivation for the reader or listener to identify the product and to take the action you are requesting. They should be written in language targeted to the intended audience.

- Letters or other communications should be specific and concise.
- The words "Important Safety Notice" or "Safety Recall" should appear at the top of each notice and cover letter and should also be on the lower left corner of any mailing envelope.
- Notices to retailers and distributors should explain the reason for the recall, including the hazard, and contain all the instructions needed to tell them how to handle their product inventory, as well as instructions for displaying posters or notices, providing information to consumers, and disposing of returned products.
- All letters and other notices to consumers should explain clearly the reason for the recall, including injury or potential injury information, and provide complete instructions.

E. Toll-Free Numbers

A company conducting a recall should provide a toll-free (800/888/877) telephone number for consumers to respond to the recall announcement. Generally, this number should be dedicated only to the recall. Historically, the Commission staff has found that most company systems for handling consumer relations or for ordering products, repairs, or accessories are unable to respond effectively to callers about recall announcements, particularly during the first few weeks after the initial announcement.

When establishing a telephone system to handle a recall, be overgenerous in estimating consumer response, especially during the first several days/weeks. It is easier to cut back than it is to add more capacity once a recall is announced, and consumers who are unable to get through may not keep trying.

Whether you use an automated system or live operators to answer the calls, prepare scripts and instructions for responding to questions. Operators or taped messages should begin by identifying the firm and product and explaining the reason for the recall. Most consumers who hear about a recall by radio, television, or word of mouth will not remember all the information they initially heard. Again, at its beginning, the message should reinforce the need for listeners to act, particularly if the message is lengthy. CPSC Compliance staff needs to review all scripts before the recall is announced. All automated systems should provide a number for consumers to contact the firm for special problems, e.g., problems completing repairs or installing parts.

VII. Developing a Company Policy and Plan to Identify Defective Products and to Undertake a Product Recall

Companies whose products come under the jurisdiction of the CPSC should consider developing an organizational policy and plan of action if a product recall or similar action becomes necessary, whether it involves the CPSC or another government agency. This policy and any related plans should focus on the early detection of product safety problems and prompt response.

A. Designating a Recall Coordinator

Designating a company official or employee to serve as a "recall coordinator" is a significant step that a firm can take to meet its product safety and defect reporting responsibilities. Ideally, this coordinator has full authority to take the steps necessary (including reporting to the Commission) to initiate and implement all recalls, with the approval and support of the firm's chief executive officer.

The recall coordinator should have the following qualifications and duties:

- Knowledge of the statutory authority and recall procedures of the Consumer Product Safety Commission;
- Ability and authority to function as the central coordinator within the company for receiving and processing all information regarding the safety of the firm's products. Such information includes, e.g., quality control records, engineering analyses, test results, consumer complaints, warranty returns or claims, lawsuits, and insurance claims.
- Responsibility for keeping the company's chief executive officer informed about reporting requirements and all safety problems or potential problems that could lead to product recalls;
- Responsibility for making decisions about initiating product recalls;
- Authority to involve appropriate departments and offices of the firm in implementing a product recall;
- Responsibility for serving as the company's primary liaison person with CPSC.

B. Role of the Recall Coordinator

At the outset, the recall coordinator should fully review the company's product line to determine how each product will perform and fail under conditions of proper use and reasonably foreseeable misuse or abuse. Through research and analysis, product safety engineers can identify the safety features that could be incorporated into products that present safety risks to reduce their potential for future injury.

The company should institute a product identification system if one is not now in use. Model designations and date-of-manufacture codes should be used on all products, whether they carry the company's name or are privately labeled for other firms. If a product recall is necessary, this practice allows the company to identify easily all affected products without undertaking a costly recall of the entire production. Similarly, once a specific product has been recalled and corrected, a new model number or other means of identification used on new corrected products allows distributors, retailers, and consumers to distinguish products subject to recall from the new items. Until a production change can be made to incorporate a new model number or date code, some companies have used sticker labels to differentiate products that have been checked and corrected from recalled products.

VIII. Records Maintenance

The goal of any product recall is to retrieve, repair, or replace those products already in consumers' hands as well as those in the distribution chain. Maintaining accurate records about the design, production, distribution, and marketing of each product for the duration of its expected life is essential for a company to conduct an effective, economical product recall. Generally, the following records are key both to identifying product defects and conducting recalls:

A. **Records of complaints, warranty returns, insurance claims, and lawsuits.** These types of information often highlight or provide early notice of safety problems that may become widespread in the future.
B. **Production records.** Accurate data should be kept on all production runs—the lot numbers and product codes associated with each run, the volume of units manufactured, component parts or substitutes use, and other pertinent information which will help the company identify defective products or components quickly.
C. **Distribution records.** Data should be maintained as to the location of each product by product line, production run, quantity shipped or sold, dates of delivery, and destinations.
D. **Quality control records.** Documenting the results of quality control testing and evaluation associated with each production run often helps companies identify possible flaws in the design or

production of the product. It also aids the firm in charting and sometimes limiting the scope of a corrective action plan.

E. **Product registration cards.** Product registration cards for purchasers of products to fill out and return are an effective tool to identify owners of recalled products. The easier it is for consumers to fill out and return these cards, the greater the likelihood the cards will be returned to the manufacturer. For example, some firms provide pre-addressed, postage-paid registration cards that already have product identification information, e.g., model number, style number, special features, printed on the card. Providing an incentive can also increase the return rate. Incentives can be coupons towards the purchase of other products sold by the firm, free accessory products, or entry in a periodic drawing for a product give away. The information from the cards then needs to be maintained in a readily retrievable database for use in the event a recall becomes necessary.

IX. Conclusion

Consumers no longer view product recalls in a negative light. Millions of products have been recalled over the years. Today, consumers believe they enjoy a safer, better product as a result of a recall conducted responsibly by company. How well a company conducts a timely, reasonable recall of a product can have a strong influence on consumers' attitude about the firm. Successful product recalls in the past have rewarded companies with continuing consumer support and demand for the firms' products.

For additional information about product recalls and reporting, call (301) 504-7913, fax (301) 504-0359, or visit the Commission's World Wide Web site at www.cpsc.gov (press the Business icon).

Bibliography

Engineering and Science References

Ashby, Michael F.; and Jones, David R.H. (1980). *Engineering Materials 1: An Introduction to Their Properties and Applications*. Oxford: Pergamon Press.

Avallone, Eugene A.; and Baumeister III, Theodore (1996). *Mark's Standard Handbook for Mechanical Engineers*. New York: McGraw-Hill.

Douglas Frink, L.J.; and van Swol, F. (1997). "A molecular theory for surface forces adhesion measurements," *J. Chem. Phys.*, vol. 106, pp.

Fischer, Martin A. (1983). *Engineering Specifications Writing Guide*. Upper Saddle River, NJ: Prentice Hall.

Frost, Harold J.; and Ashby, Michael F. (1982). *Deformation-Mechanism Maps: The Plasticity and Creep of Metals and Ceramics*. Oxford: Pergamon Press.

Harper, Charles A. (2001). *Handbook of Materials for Product Design*. New York: McGraw-Hill.

Oberg, E.; Jones, F.D.; and Horton, H.L. (2004). *Machinery's Handbook*, 27th edition. New York: Industrial Press.

McCrum, N.G.; Buckley, C.P.; and Bucknall, C.B. (2003). *Principles of Polymer Engineering*. New York: Oxford Science Publications.

Moad, G. (1992). *Polymer Bulletin*, vol. 29, pp. 647–652.

Norton, Robert L. (2008). *Design of Machinery*. New York: McGraw-Hill.

Park, Chan S. (2007). *Contemporary Engineering Economics*, 3rd edition. Upper Saddle River, NJ: Prentice Hall.

Rasis, E.P. (1991). *Technical Reference Handbook*. Orland Park, IL: American Technical Publishers.

Shigley, J.E.; and Mischke, C.R. (1989). *Mechnical Engineering Design*, 5th edition. New York: McGraw-Hill.

Sullivan, William G.; Wicks, Elin M.; and Luxhoj, James T. (2008). *Engineering Economy*, 30th edition. Upper Saddle River, NJ: Prentice Hall.

Turner, S. (2001). "Creep of Polymeric Materials." *Encyclopedia of Materials: Science and Technology*. Oxford: Elsevier Science.

Failure Analysis References

Brooks, Charlie R.; and Choudhury, Ashok. (2002). *Failure Analysis of Engineering Materials*. New York: McGraw-Hill.

Brostow, Witold; and Corneliussen, Roger D. (1989). *Failure of Plastics*. Munich: Hanser Publishers.

Carlson, R.L.; and Kardomateas, G.A. (1996). *An Introduction to Fatigue in Metals and Composites*. London: Chapman and Hall.

Collins, Jack A. (1993). *Failure of Materials in Mechanical Design—Analysis, Prediction, Prevention*. New York: John Wiley and Sons.

Goodell, Barry; Nicholas, Darrel D.; and Schultz, Tor P. (2003). *Wood Deterioration and Preservation*, ACS Symposium Series 845.

Wulpi, Donold J. (1999). *Understanding How Components Fail*, 2nd edition. Materials Park, OH: ASM International.

Purchasing and Sourcing References

Bower, Edward K. (2003). *Specification-Driven Product Development*. Bloomington, IN: iUniverse.

Campbell, John D.; and Reyes-Picknell, James V. (2006). *Uptime—Strategies for Excellence in Maintenance Management*, 2nd edition. New York: Productivity Press.

Colton, Raymond R.; and Rohrs, Walter F. (1985). *Industrial Purchasing and Effective Materials Management*. Reston, VA: Reston Publishing.

DePaoli, Tom. (2004). *Common Sense Purchasing*. Charleston, SC: BookSurge.

Feron, Harold E.; Dodler, Donald W.; and Killen, Kenneth H. (1989). *The Purchasing Handbook*. New York: McGraw-Hill.

Grieco, Peter L. (1997). *MRO Purchasing*. West Palm Beach, FL: PT Publications.

Gryna, F.M.; Chua, R.C.H.; and Defeo, J.A. (2002). *Juran's Quality Planning and Analysis for Enterprise Quality*, 5th edition. New York: McGraw-Hill.

Hough, Harry E. (1996). *Purchasing for Manufacturing*. New York: Industrial Press.

Johnson, Nick. "Interesting Facts on Samurai Sword Manufacture," http://ezinearticles.com/?Interesting-Facts-on-Samurai-Sword-Manufacture&id=61344.

Howe, Irving, ed. (1982). *The Portable Kipling*. London: Penguin Books.

King, Donald B.; and Ritterskamp, James R. Jr. (1993). *Purchasing Manager's Desk Book of Purchasing Law*, 2nd edition. Upper Saddle River, NJ: Prentice Hall.

Laseter, Timothy M. (1998). *Balanced Sourcing*. San Francisco: Jossey-Bass.

Maas, Richard A.; Brown, John O.; and Bossert, James L. (1990). *Supplier Certification—A Continuous Improvement Strategy*. Seattle, WA: Quality Press.

Milwaukee Power Tool Battery. *Recall: July 10, 2007*, U.S. Consumer Product Safety Commission, Release #07-234, http://www.cpsc.gov/cpscpub/prerel/prhtml07/07234.html.

Purdy, David C. (1993). *A Guide to Writing Successful Engineering Specifications*. New York: McGraw-Hill.

Rogers, Elizabeth; and Kostigen, Thomas M. (2007). *The Green Book*. New York: Three Rivers Press.

Walker, Ian. (May 29, 2002). *Time Is Money*, CNN.Com, http://articles.cnn.com/2002-05-29/tech/time.money_1_pence-formula-average-cost?_s=PM:TECH.

Index

"f" indicates material in figures; "g" indicates material in graphs; "n" indicates material in footnotes; "t" indicates material in tables.

B

D

F

G

J

K

O

P

Q

R

S

T

U

X

Y

Z

For Product Safety Concerns and Information please contact our EU representative GPSR@taylorandfrancis.com Taylor & Francis Verlag GmbH, Kaufingerstraße 24, 80331 München, Germany

Printed and bound by CPI Group (UK) Ltd, Croydon, CR0 4YY
08/05/2025
01864394-0006